D1237287

WILEY SERIES IN PROBABILITY
AND MATHEMATICAL STATISTICS

ESTABLISHED BY WALTER A. SHEWHART AND SAMUEL S. WILKS
Editors
Vic Barnett, Ralph A. Bradley, J. Stuart Hunter,
David G. Kendall, Rupert G. Miller, Jr., Adrian F. M. Smith,
Stephen M. Stigler, Geoffrey S. Watson

Probability and Mathematical Statistics

(*continued on back*)

Multivariate Statistical Simulation

Multivariate Statistical Simulation

MARK E. JOHNSON

Statistics and Operations Research
Los Alamos National Laboratory
Los Alamos, New Mexico

JOHN WILEY & SONS

New York · Chichester · Brisbane · Toronto · Singapore

A NOTE TO THE READER
This book has been electronically reproduced from
digital information stored at John Wiley & Sons, Inc.
We are pleased that the use of this new technology
will enable us to keep works of enduring scholarly
value in print as long as there is a reasonable demand
for them. The content of this book is identical to
previous printings.

Copyright © 1987 by John Wiley & Sons, Inc.

All rights reserved. Published simultaneously in Canada.

Reproduction or translation of any part of this work
beyond that permitted by Section 107 or 108 of the
1976 United States Copyright Act without the permission
of the copyright owner is unlawful. Requests for
permission or further information should be addressed to
the Permissions Department, John Wiley & Sons, Inc.

Library of Congress Cataloging in Publication Data:

Johnson, Mark E. 1952–
    Multivariate statistical simulation.

    (Wiley series in probability and mathematical
statistics. Applied probability and statistics)
    Includes bibliographies and index.
    1. Multivariate analysis—Data processing.   I. Title.
II. Series.

QA278.J62 1987   519.5′35′0724   86-22469
ISBN 0-471-82290-6

Printed in the United States of America

10 9 8 7 6 5 4 3 2

Preface

Multivariate Statistical Simulation concerns the computer generation of multivariate probability distributions. Generation is used in a broader context than solely algorithm development. An important aspect of generating multivariate probability distributions is what should be generated, as opposed merely to what could be generated. This viewpoint necessitates an examination of distributional properties and of the potential payoffs of including particular distributions in simulation studies. Since other available books document many of the mathematical properties of distributions (e.g., Johnson and Kotz, *Distributions in Statistics: Continuous Multivariate Distributions*), a complementary approach is taken here. Generation algorithms are presented in tandem with many graphic aids (three-dimensional and contour plots) that highlight distributional properties from a unique perspective. These plots reveal features of distributions that rarely emerge from preliminary algebraic manipulations.

The primary beneficiary of this book is the researcher who is confronted with the task of designing and executing a simulation study that will employ continuous multivariate distributions. The prerequisite for the reader is a relentless curiosity as to the behavior of the method, estimator, test, or system under investigation when various multivariate distributions are assumed. The multivariate distributions presented in this text can serve as simulation drones to satiate the researcher's curiosity.

For the past ten years or so, my research efforts have consistently involved the development of new distributions to be used in simulation contexts. Hence several chapters reflect my naturally biased disposition toward certain distributions (Pearson Types II and VII elliptically contoured distributions, Khintchine distributions, the unifying class for the Burr, Pareto, logistic distributions). As a reasonable attempt for completeness, various multivariate distributions that are potential (but as yet not sufficiently developed) competitors to the highlighted distributions are mentioned in the research directions chapter or in the supplementary bibliography.

Although not designed as a text, this book can be used as the primary reference in a graduate seminar in simulation. Exercises could consist of adapting for simulation purposes various references in the supplementary list.

The initial draft of this book was written while I was on sabbatical at the University of Arizona and the University of Minnesota during the academic year 1982–1983. For the Tucson connection I am grateful to John Ramberg, Chairman of the Systems and Industrial Engineering Department, and to Chiang Wang, who gave me considerable support. The Minnesota visit was made possible by the efforts of Dennis Cook, Chairman of the Department of Applied Statistics, and financial support was provided by the School of Statistics under the aegis of Seymour Geisser. The hospitality at both departments is gratefully acknowledged. Valuable insights were provided by Dick Beckman (Los Alamos), Christopher Bingham (St. Paul), Adrian Raftery (Seattle), and George Shantikumar (Berkeley). I am particularly indebted to Sandy Weisberg at Minnesota, whose careful reading of the manuscript led to significant improvements. I am grateful to Myrle Johnson and Geralyn Hemphill for computer graphics support. Finally, this book would not have been possible without the continued support of Larry Booth and Harry Martz, Jr. at Los Alamos.

Skilled typing was performed by Kathy Leis and Kay Woefle (Tucson), Carol Leib and Terry Heineman-Baker (St. Paul), and Kay Grady, Hazel Kutac, Sarah Martinez, Corine Ortiz, and Esther Trujillo (Los Alamos).

MARK E. JOHNSON

Los Alamos, New Mexico
October 1986

Contents

CHAPTER 1

Introduction

Monte Carlo methods are becoming widely applied in the course of statistical research. This is particularly true in small-sample studies in which statistical techniques can be scrutinized under diverse settings. Developments in computing have also encouraged the creation of new methods, such as bootstrapping (Efron, 1979), which exploit this capability. In these respects, statistical research and computing have evolved a symbiotic relationship.

Monte Carlo studies as reported in the statistical literature typically result from the progression of tasks outlined in Figure 1.1. A new statistical technique is first conceived and its associated properties are sought. The main goal of the investigation is probably to collect evidence that will persuade others to employ the method. There are bound to be some characteristics of the new method that resist mathematical analysis, in which case Monte Carlo methods may be used to provide additional knowledge. A preliminary or pilot Monte Carlo study might detect any obvious flaws or possible improvements in the new procedure or suggest numerically efficient shortcuts. Next a larger scale Monte Carlo study is designed in order to address the open questions about the method's properties. A key step in this particular task is the selection of cases, which involves the choice of distributions and their parameters, sample sizes, and so forth. The large-scale study is then conducted and the results synthesized. In the happy situation that the new method performs "well," the investigator can proceed to the publication stage. Otherwise, adjustments to the simulation design or the method itself may be pondered and various tasks repeated.

Some possible purposes of the generic study described in Figure 1.1 might include examination of robustness properties, assessment of small-sample versus asymptotic agreement, or comparison of the new method

1

with its competitors. This flow chart is generally appropriate for studies that involve either univariate or multivariate distributions. However, the problems in the design stages are vastly different with regard to case selection. For a study in which univariate distributions are used, the problem is to select from among the many distributions available. The set of continuous univariate distributions that can fairly easily be used includes the following:

Beta	Kappa
Burr	Lambda
Cauchy	Laplace
Contaminated normal	Logistic
Exponential power	Normal
Extreme value	Pareto
F	Pearson system
Gamma (including χ^2 and exponential)	Slash
Generalized gamma	Stable
Inverse Gaussian	t
Johnson system (including lognormal)	Weibull

With relative ease, an investigator can accumulate vast quantities of numerical results. However, broad coverage of distributions garnered from the extensive use of the above list can make the subsequent assimilation of results difficult. In particular guiding principles can be lacking as to the effect of distribution on the statistical method under study. Some authors such as Pearson, D'Agostino, and Bowman (1977) have resorted to tabulating distributional results according to the population skewness and kurtosis values. This tactic provides at best a crude ordering for diverse univariate distributions.

In contrast to the univariate setting in which many distributions are available, the multivariate setting offers relatively few distributions that are suitable in Monte Carlo contexts. Although there are many multivariate distributions—the texts by Johnson and Kotz (1972) and Mardia (1970) attest to this—the key word is suitable. More recent advances, as can be found in the NATO conference volumes following international meetings in Calgary (Patil, Kotz, and Ord, 1975) and Trieste (Taillie, Patil, and Baldessari, 1981), tend to be inappropriate or incomplete for application in Monte Carlo studies. Some current limitations of many of these multivariate distributions with respect to Monte Carlo work include the following:

1. Many distributions are tied directly to sampling distributions of statistics from the usual multivariate normal distribution. Outside this

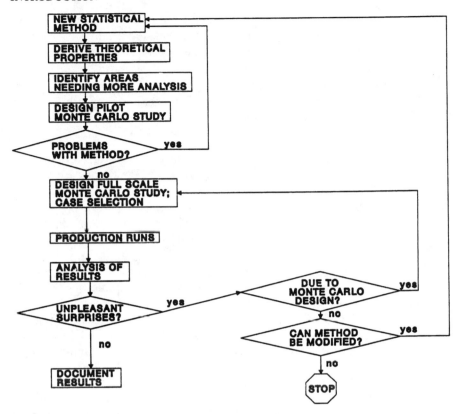

Figure 1.1. Generic Monte Carlo study.

realm of normal theory inference or estimation, the distributions may have little to offer.

2. Other distributions present formidable computational problems (e.g., Bessel function distributions).

3. The support of some of the distributions is too restrictive to be of general interest. Possible examples include the beta-Stacy distribution (Mihram and Hultquist, 1967) and some of the distributions developed by Kimeldorf and Sampson (1978).

4. Some distributions are limited to modeling only weak dependence. Morgenstern's distribution (Section 10.1) is such an example, since by any measure of association, its intrinsic dependence is restrictive. Also, the trivial case of multivariate distributions constructed with independent uni-

variate components has this obvious shortcoming. Multivariate distributions with independent components are, however, important as a baseline for assessing the effects of nontrivial dependence. This issue is explored in detail in some specific contexts later.

5. Computational support for some distributions is lacking. For example, no method may have been published for generating variates from the distribution, or if a method is known, the required univariate generation routines are unavailable.

These limitations of some existing distributions should not be viewed as grounds to abandon those distributions entirely. The limitations are cited to explain their rare use, which might be remedied given particular advances in research. Morgenstern's distribution, which has limited dependence structure in its own right, can be incorporated neatly with the Burr, Pareto, and logistic distribution of Section 9.1 to yield a valuable general distribution developed in Section 9.2.

Deficiencies in currently available distributions further point to the general issue concerning the purposes of Monte Carlo studies and the role of multivariate distribution selection to accommodate these purposes. In the absence of a particular investigation, general recommendations for distribution selection are difficult to provide. Most new statistical techniques have some basic characteristics or nuances that can influence case selection and the design of the Monte Carlo study. To illustrate this point, three distinct research topics are outlined in Sections 1.2–1.4. These discussions are intended to illustrate the potential benefits of problem analysis prior to the execution of the study. In addition these sections provide more justification for the developments given in subsequent chapters. The three topics are the robustness of Hotelling's T^2-test (Everitt, 1979), error rates in partial discriminant analysis (Beckman and Johnson, 1981), and a new multivariate goodness-of-fit test (Foutz, 1980).

These topics are described in some detail, as they may be of independent interest and they may provide an appreciation for the problems and challenges awaiting future investigations and developments in Monte Carlo studies. These are certainly the types of studies that have motivated the writing of this text and have influenced the coverage of distributions given in subsequent chapters.

1.1. ROBUSTNESS OF HOTELLING'S T^2 STATISTIC

A standard problem in multivariate analysis is to test the equality of an unknown population mean vector μ and a specified mean vector μ_0. This test can be conducted using a random sample $X_1, X_2, \ldots, X_n$ from a multivariate normal distribution denoted $N_p(\mu, \Sigma)$ where p is the dimension and Σ is a $p \times p$ covariance matrix. The appropriate test statistic is Hotelling's T^2, computed as

$$T^2 = n(\overline{\mathbf{X}} - \mu_0)' S^{-1} (\overline{\mathbf{X}} - \mu_0),$$

where

$$\overline{\mathbf{X}} = \frac{1}{n} \sum_{i=1}^{n} \mathbf{X}_i, \quad \text{and} \quad S = \frac{1}{n-1} \sum_{i=1}^{n} (\mathbf{X}_i - \overline{\mathbf{X}})(\mathbf{X}_i - \overline{\mathbf{X}})'.$$

Under the null hypothesis that $\mu = \mu_0$, the statistic $(n - p)T^2/p(n - 1)$ has an F distribution with p and $n - p$ degrees of freedom. Hotelling's T^2 is the basis of a uniformly most powerful test against the alternative $\mu \neq \mu_0$ and is invariant under nonsingular linear transformation (Muirhead, 1982, pp. 211–215). In simple terms, if the assumptions are valid that the $\mathbf{X}_i$'s are independent and identically distributed $N_p(\mu, \Sigma)$, then T^2 should be used for inference.

An important practical issue concerns the performance of T^2 if the underlying assumptions are incorrect. There are a variety of ways in which the assumptions could go awry, only a few of which have been addressed in the literature (Everitt, 1979 and Nath and Duran, 1983). The typical situation involves $\mathbf{X}_i$'s that have independent and identically distributed components following a simple univariate distribution such as rectangular, exponential, or lognormal. Of course, the normal distribution is usually included in these studies as a check on the computer program. Alternatives with independent components are not so restrictive as might be surmised, since the results for a given set of $\mathbf{X}_i$'s would be identical as those for $A\mathbf{X}_i$, $i = 1, 2, \ldots, n$, for a nonsingular $p \times p$ matrix A. However, it is not sufficient to consider only random vectors with independent components. Some distributions such as the multivariate Pearson II and VII distributions (Section 6.2) cannot be obtained in this manner. The following comments outline possible areas of research on the performance of T^2 when the assumptions are violated. Some of these issues can be handled readily with the distributions described in later chapters of this book.

1. Suppose the assumption of independence in the random sample is invalid? This would mean the $\mathbf{X}_1, \mathbf{X}_2, \ldots, \mathbf{X}_n$ could be thought of as one realization from an $n \times p$ dimensional multivariate distribution. How does this affect T^2?

2. Using the independent components model for the $\mathbf{X}_i$'s, what sort of problems arise if the components are not identically distributed?

3. Moreover, is it really reasonable to assess the effects of the component distributions in terms of the univariate population skewness and kurtosis

values, as has been done by Everitt, for example? Is it possible to isolate these effects to avoid the usual confounding? Everitt demonstrated some cases in which lognormal components evidently degraded T^2 performance more than exponential components. Since the skewness of the lognormal is greater than the skewness of the exponential, he argued that skewness was the culprit. However, this argument can as well be applied using kurtosis instead of skewness, so that a more controlled experiment seems warranted.

4. For extremely non-normal cases that give terrible results with T^2, the non-normality is probably apparent, in which case a transformation to normality could be sought. Can this idea be used to ameliorate the performance of T^2?

5. The performance of T^2 has been viewed primarily in terms of holding the α-level or Type I error rate at a nominal prespecified value. A possible extension is that in cases where this robustness holds, what effects, if any, can be observed in terms of power?

6. Consideration of dimension and sample size are critical for any of the above topics. The general question related to each of these two factors is: Do the results improve, degrade, or remain unaffected as these factors vary?

With this vast set of factors of interest, Monte Carlo experiments obviously should not be conducted without considerable planning. Relatively little attention has been given to experimental design principles in the context of Monte Carlo studies, although there are exceptions (Margolin and Shruben, 1978). Additional work in this area would be welcomed.

1.2. ERROR RATES IN PARTIAL DISCRIMINANT ANALYSIS

Discriminant analysis involves techniques for classifying individuals into one of several populations on the basis of vectors of observations taken on the individuals and on the constituents from each population. Many discriminant analysis techniques implicitly assume *forced* discrimination, in which every "new" individual is to be classified. Broffitt, Randles, and Hogg (1976) described a method for partial discrimination in which an additional option—do not classify—is allowed. Subsequently, Beckman and Johnson (1981) advocated a related partial discriminant analysis method appropriate in the two-population case. This method is first described briefly and then its performance from previous Monte Carlo work is surveyed. In keeping with the spirit of the Hotelling's T^2 example above, a number of research questions are then posed. Some of these questions could be addressed through Monte Carlo studies employing the techniques and the multivariate distributions given in this book.

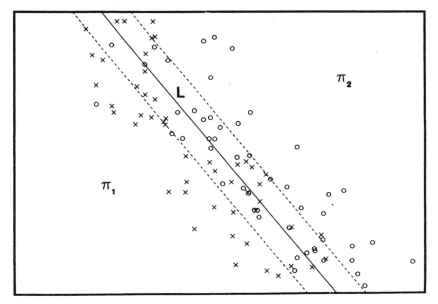

Figure 1.2. Forced and partial discrimination.

Figure 1.2 is useful for describing a simple discriminant rule for partial classification in bivariate populations. The available data on the two populations, denoted π_1 and π_2, is represented by the X's and O's in the figure. In forced discrimination, a new observation Z would be classified in π_1, for example, if it were to the left of the solid line L, and in π_2 otherwise. It should be apparent that many errors in classification will be made because of the considerable overlap in the populations. On the other hand, with partial discriminant analysis, a third region—the do-not-classify area —is included; it is the area enclosed by the dashed lines. Only new observations outside this region will be classified, and then presumably with a high probability of success.

To automate classification, it is convenient to assign a univariate score to each observation. Thus classification decisions can be made by considering certain intervals of the real line, as can be seen from the figure. The scoring function used there is the linear discriminant function given by the projection

$$L(\mathbf{z}) = (\overline{\mathbf{X}} - \overline{\mathbf{Y}})S^{-1}\mathbf{z},$$

where $\overline{\mathbf{X}}$ and $\overline{\mathbf{Y}}$ are the sample mean vectors from the training sets of observations and S is an estimated pooled covariance matrix. In particular, if the training sets are given by $T_1 = \{X_1, X_2, \ldots, X_{n_1}\}$ for π_1 and $T_2 = \{Y_1, Y_2, \ldots, Y_{n_2}\}$ for π_2, then

$$\overline{X} = \frac{1}{n_1} \sum_{i=1}^{n_1} X_i$$

$$\overline{Y} = \frac{1}{n_2} \sum_{i=1}^{n_2} Y_i$$

$$S = \frac{1}{n_1 + n_2 - 2} \left[\sum_{i=1}^{n_1} (X_i - \overline{X})(X_i - \overline{X})' + \sum_{i=1}^{n_2} (Y_i - \overline{Y})(Y_i - \overline{Y})' \right].$$

Having assigned scores via this or most any other reasonable scheme, the next task is to determine the do-not-classify interval endpoints a and b. If one knew the probability distribution of the scores from each population, represented by the densities f_1 and f_2, then the following optimization problem would need to be solved:

$$\underset{a,\, b}{\text{maximize}} \ F_1(a) + 1 - F_1(b) + F_2(a) + 1 - F_2(b) \qquad (1.1)$$

such that

$$\frac{1 - F_1(b)}{1 - F_1(b) + F_1(a)} \leqslant \alpha_1 \qquad (1.2)$$

$$\frac{F_2(a)}{F_2(a) + 1 - F_2(b)} \leqslant \alpha_2, \qquad (1.3)$$

where α_i is the specified probability of misclassification of individuals from population i given an attempted classification and F_i is the distribution function corresponding to f_i. The objective function corresponds to the proportion of observations classified. The constraints (1.2) and (1.3) relate to attaining a specified conditional error rate. In any realistic case, the F_i's are unknown but can be estimated by the sample distribution function of scores. The above optimization problem then becomes discrete and can be solved by simple enumeration. An example is provided in Table 1.1. Fifteen observations from two populations were assigned scores, and the nominal error rates are specified as $\alpha_1 = \alpha_2 = 0.10$. The four possible locations for a or b are selected from $\{c_1, c_2, c_3, c_4\}$. Any other choice, such as c_i between the scores of two X's, for example, could be improved. For each of the ten

TABLE 1.1.

Ordering of Scores

XXXXXXXXX O X OOO XXXX OO X OOOOOOOOO

$\quad\quad\quad\uparrow\quad\quad\uparrow\quad\quad\quad\quad\uparrow\quad\quad\quad\uparrow$

$\quad\quad\quad c_1\quad\quad c_2\quad\quad\quad\quad c_3\quad\quad\quad c_4$

Enumeration of Possible Solutions

a	b	Number of X's Incorrect / Number of X's Classified	Number of O's Incorrect / Number of O's Classified	Total Number Classified
c_1	c_1	6/15	$0/15^a$	30
c_1	c_2	5/14	$0/14^a$	28
c_1	c_3	$1/10^a$	$0/11^a$	21^b
c_1	c_4	$0/9^a$	$0/9^a$	18^b
c_2	c_2	5/15	$1/15^a$	30
c_2	c_3	$1/11^a$	$1/12^a$	23^b
c_2	c_4	$0/10^a$	$1/10^a$	20^b
c_3	c_3	$1/15^a$	4/15	30
c_3	c_4	$0/14^a$	4/13	27
c_4	c_4	$0/15^a$	6/15	30

[a] Observed conditional error rates less than or equal to 0.10.
[b] Both conditional error rates acceptable.

possible pairs (a, b), the error rates on the training sets themselves are computed. Four of the ten cases, namely (c_1, c_3), (c_1, c_4), (c_2, c_3), and (c_2, c_4), attain the specified 10% error rate in each population, at least on the training sets. Of these, the pair (c_2, c_3) classifies the most number of individuals, 23.

One additional adjustment to this method is necessary to ensure its decent performance in small samples. The reason for an adjustment is that discrimination performance is overly optimistic on the training sets in comparison to new observations. A computing-intensive scheme for reducing this bias was developed by Beckman and Johnson (1981), and a brief description is included here for completeness. Assume a scoring function L whose parameters can be estimated from the training sets T_1 and T_2, as given above. A new observation $\mathbf{Z}$ is to be classified. We first pretend that $\mathbf{Z}$ belongs to population π_1 so that our training sets are $\{\mathbf{Z}\} \cup T_1$ and T_2 of size $n_1 + 1$ and n_2, respectively. A scoring function L_1 is then determined based on these training sets. The discrete optimization problem represented by (1.1)–(1.3) is solved to obtain cutoff points a_1 and b_1 and pertinent intervals $A_1 = (-\infty, a_1]$ and $B_1 = [b_1, \infty)$. Clearly, $L_1(\mathbf{Z}) \in A_2$ is evidence to suspect that Z belongs to π_1 and $L_1(Z) \in B_1$ suggests that Z belongs to π_2. The previous steps are then repeated pretending that Z belongs to π_2. A revised scoring function L_2 based on training sets T_1 and $T_2 \cup \{Z\}$ is determined and new cutoff values a_2 and b_2 are calculated leading to intervals $A_2 = (-\infty, a_2]$ and $B_2 = [b_2, \infty)$. If $L_2(Z) \in A_2$, then perhaps Z belongs to π_1; if $L_2(Z) \in B_2$, then possibly $Z \in \pi_2$. The results of these two exercises lead to the following classification rule:

If $L_1(Z) \in A_1$ and $L_2(Z) \in A_2$, then classify Z in π_1.
If $L_2(Z) \in B_1$ and $L_2(Z) \in B_2$, then classify Z in π_2.
Otherwise, do not classify Z.

This classification rule seems to be asymptotically distribution-free although explicit restrictions on the distributions that underly populations π_1 and π_2 and the scoring functions have not been derived. However, some of the small-sample properties have been considered and no evidence to refute this point has emerged. A brief review of the cases examined by Beckman and Johnson is now given. Of particular interest was the performance of this method for a variety of distributions. Three bivariate distributions for the populations were considered: normal (Section 4.1), t (Section 6.2), and lognormal (Chapter 5). For each distributional case, four subcases set the population covariance matrices equal, $\Sigma_1 = \Sigma_2$ and four subcases have $\Sigma_1 \neq \Sigma_2$. The $\Sigma_1 = \Sigma_2$ cases considered correlations between the two components as 0, $\frac{1}{4}$, $\frac{1}{2}$, and $\frac{3}{4}$. In the $\Sigma_1 \neq \Sigma_2$ case, population π_1 was governed by independence ($\Sigma_1 = I$) and in π_2, Σ_2 had four possibilities obtained from two choices of ρ($\frac{1}{4}$ or $\frac{3}{4}$) crossed with two choices of (σ_1, σ_2) —either $(1, 1)$ or $(1, 2)$. Finally, sample sizes were varied as 21, 35, 51, 101, 201. Thus there were 120 individual cases considered although only three specific distributions were used. This may be a typical phenomenon—rarely can "distribution" be treated as an isolated treatment in the design sense. Sample size is important in general and covariance structure is particularly important in discriminant analysis studies. The results of the study basically indicated that for sample sizes 51 and larger, the proposed partial discriminant analysis method worked correctly—the nominal conditional error rates were achieved.

The results of this Monte Carlo study were not intended to be extrapolated to cover all multivariate distributions, covariance structures, and their combinations with two populations. The results did, however, appear sufficiently promising to apply the technique in a geological investigation that had previously provided tacit inspiration for the cases in the Monte

Carlo study (Patterson et al., 1981). In fact, there are many additional avenues of investigation that could be pursued, some of which are given below:

1. Does the procedure continue to perform correctly in higher dimensions? It might be surmised that larger sample sizes would be required to attain comparable results.

2. In the previous studies, the distributional forms underlying π_1 and π_2 were the same—only the parameters varied. Are the results different if π_1 and π_2 originate from distinct distributions? For example, suppose π_1 is governed by a normal and π_2 by a lognormal?

3. For cases in which π_1 and π_2 have the *same* mean vector, discrimination could be made on the basis of dispersion about the mean. Assuming adjustments to the basic partial discriminant analysis method can be discerned, what sort of performance is achievable?

4. Certainly other scoring functions could be considered. Also, different estimators of the parameters of a particular scoring function could be tried. For example, robust estimates of the μ's and Σ could be calculated. Is there an increase in the proportion of observations classified and if so, is it sufficient to justify the additional effort in calculating these estimates? Is it possible to recommend particular scoring functions for various classes of distributions?

5. Extensive Monte Carlo studies provide some assurance that the method applied to real data will be satisfactory. A practitioner may want additional guarantees that on the specific data being considered the specified conditional error rates will really be achieved and the maximum number of individuals will be classified. One possible approach to collecting this evidence is to apply Efron's (1979) bootstrapping technique to estimate the standard errors of the estimated conditional error rates and proportion of classifications Efron gives an example of the bootstrap's use in forced discrimination so that only a slight adaptation is necessary for the partial discrimination problem.

1.3. FOUTZ' TEST

Foutz (1980) developed a new general-purpose goodness-of-fit test that can be applied in multivariate situations. This test has a number of intuitively appealing features that encourage a thorough empirical examination. First Foutz' test is described and then, in keeping with the previous two examples, a set of research questions is posed. The primary intent is to provide additional motivation for having included the material in subsequent chapters.

Suppose $X_1, X_2, \ldots, X_{n-1}$ constitute a random sample distributed according to a probability measure P that is assumed to be absolutely continuous with respect to Lebesque measure. The problem is to test the hypothesis that $P = P_0$ where P_0 is a specified probability measure. In comparing these two measures, it is natural to find the Borel set in R^P, say B^*, for which $|P(B^*) - P_0(B^*)|$ is a maximum. This is a tricky problem to say the least—searching through the Borel sets. Foutz devises an ingenious scheme for conducting a test of $P = P_0$, a scheme that requires the computation of probabilities of only n Borel sets. These sets are generated through the construction of statistically equivalent blocks $B_1, B_2, \ldots, B_n$, which depend on the data in a manner to be described shortly. Given these blocks and the hypothesized measure P_0, Foutz' test statistic F_n is computed as follows:

1. Compute $D_i = P_0(B_i)$, $i = 1, 2, \ldots, n$.
2. Order these probabilities as $D_{(1)} < D_{(2)} < \cdots < D_{(n)}$.
3. Evaluate $F_n = \max_{j=1,2,\ldots,n-1}(j/n - D_{(1)} - D_{(2)} - \cdots - D_{(j)})$.

For very small values of n (< 5) the null distribution of F_n can be worked out exactly. For intermediate values up to possibly 50, a Monte Carlo approach can be used. Franke and Jayachandran (1983) report critical values for $n = 20, 30, 50$ at the significance levels 0.10, 0.05, and 0.01. Asymptotically, F_n is normally distributed with mean e^{-1} and variance $(2e^{-1} - 5e^{-2})/n$. Of particular importance is that the distribution of F_n is the same for any dimension p.

It remains to describe how the statistically equivalent blocks B_1, $B_2, \ldots, B_n$ are constructed. The method is due to Anderson (1966) and described by Foutz. Here the basic setup using Foutz' notation is described and then the mechanics are carried out on a simple example. To start, the $n - 1$ data points $X_1, X_2, \ldots, X_{n-1}$ are in R^P, which in the block notation is $B_{12\ldots n}$. Suppose there are $n - 1$ real-valued functions $h_1(x), h_2(x), \ldots,$ $h_{n-1}(x)$ such that $h_i(X)$ is a continuous random variable $i = 1, 2, \ldots, n - 1$, if X is distributed according to P.

Each function is used to identify one data point for purposes of partitioning R^P or a subset of R^P. These functions are subsequently considered in the order given by a permutation $K_1, K_2, \ldots, K_{n-1}$ of the integers $1, 2, \ldots, n - 1$. At the first stage, the function h_{K_1} is used and the values $h_{K_1}(X_1), \ldots, h_{K_1}(X_{n-1})$ are computed. The K_1th smallest such value

TABLE 1.2. Example for Constructing Some Statistically Equivalent Blocks

Data

	Cut Functions
$X_1 = (-3, 1)$	$h_i(x_1, x_2) = x_1$ if i is odd
$X_2 = (-2, 2)$	$= x_2$ if i is even
$X_3 = (3, -2)$	
$X_4 = (2, 1)$	Permutation
$X_5 = (-1, -3)$	$K_1 = 2, K_2 = 4, K_3 = 1, K_4 = 3,$
$X_6 = (1, 3)$	$K_5 = 6, K_6 = 5$

is determined; let us suppose it corresponds to $X^{(K_1)}$. The set $B_{12\ldots n}$ is now decomposed into

$$B_{12\ldots K_1} = \left\{ X: h_{K_1}(X) \leqslant h_{K_1}(X^{(K_1)}) \right\}$$

$$B_{K_1+1\ldots n} = \left\{ X: h_{K_1}(X) > h_{K_1}(X^{(K_1)}) \right\}.$$

The data point $X^{(K_1)}$ is no longer used in constructing additional blocks. At the second stage, the function h_2 is used to subdivide $B_{12\ldots K_1}$ if $K_2 < K_1$ or to subdivide $B_{K_1+1\ldots n}$ if $K_2 > K_1$. In most descriptions of such constructions, at this point the notation begins to become awkward. Thus for purposes of clarity, we switch now to a simple example. Six hypothetical bivariate data points are given in Table 1.2 along with six h_i functions and one possible permutation of $1, 2, \ldots, 6$. Figure 1.3 illustrates the construction of the blocks through the six stages. The first function considered is $h_2(x_1, x_2) = x_2$, since $k_1 = 2$. The second smallest value of $h_2(x_1, x_2)$ is -2, which corresponds to data point X_3. The boundary $y = -2$ is plotted and, of course, passes through X_3 and generates blocks B_{12} and B_{34567}. The next function considered is $h_4(x_1, x_2) = x_2$, as $k_2 = 4$. Subdivision will occur to the block B_{34567} since 4 is one of its subscripts. At the end of this stage we wish to have two new blocks B_{34} and B_{567}. To accomplish this, we evaluate h_4 at the four data points X_1, X_2, X_4, and X_6, which are in B_{34567}. The second to the largest value is 1 corresponding to data point X_4. Another boundary is drawn in and the appropriate blocks labeled. The procedure continues in this fashion through six stages and culminates with the final construction of seven blocks.

It should be noted that the construction of statistically equivalent blocks is greatly facilitated if the h_i functions are identical. In this case, the blocks are the same regardless of the permutation $K_1, K_2, \ldots, K_{n-1}$. Another consideration is to select blocks so that $P_0(B)$ can be computed without

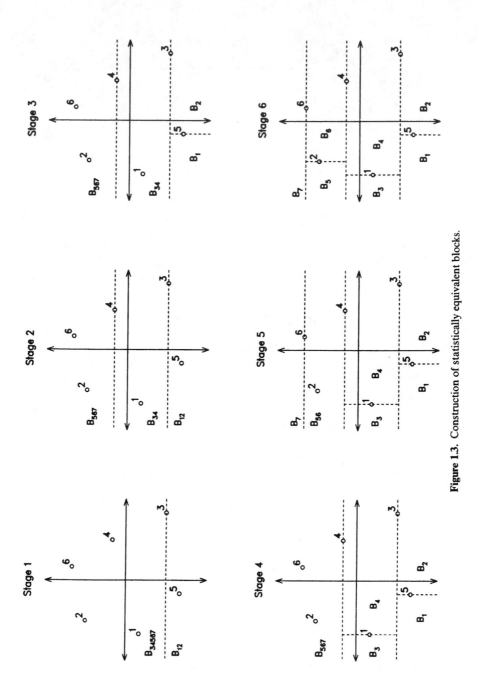

Figure 1.3. Construction of statistically equivalent blocks.

14

overwhelming difficulty. For example, in the case that P_H is the measure corresponding to a p-dimensional multivariate normal distribution with mean vector μ and covariance matrix Σ, the functions

$$h_i(\mathbf{x}) = (\mathbf{x} - \mu)'\Sigma^{-1}(\mathbf{x} - \mu), \qquad i = 1, 2, \ldots, n - 1$$

are convenient choices. For each $\mathbf{X}_i$, $h_i(\mathbf{X}_i)$ has a chi-squared distribution with p degrees of freedom.

With the basic description given above, it is now possible to identify a number of research questions that could be addressed with the results given in later chapters. The following list provides some of the preliminary queries of interest. Subsequent lines of research will, of course, depend on these outcomes.

1. How well do tests based on F_n maintain a specified α-level? Franke and Jayachandran found that critical values based on the normal approximation were conservative for $n - 1 = 20$, 30, and 50. They used a Monte Carlo approach of drawing samples of size 25,000 from F_n and then computing the appropriate percentiles. If F_n leads to a decent test, further tabulations with more sample sizes should be pursued.

2. Goodness of a test is frequently equated with power in relation to competitors. Test for multivariate normality would seem to be a natural application for Foutz' general method. In this realm, the leading competitors are union-intersection tests developed by Malkovich and Afifi (1973) and a test due to Hawkins (1981). A fundamental issue is to determine what distributional alternatives to the normal are best noticed by the various tests.

3. For Foutz' test to be of interest to practitioners, it must be adjusted to handle estimating parameters in the hypothesized distribution. An empirical examination should be performed as to the effect on the critical values if the parameters of P_0 are estimated. To what extent do the corresponding critical values depend on these unknown parameters? Is the distribution of F_n severely affected by dimension if the parameters of P_0 are estimated?

1.4. OVERVIEW

These three research topics—robustness of T^2, error rates in partial discrimination, and Foutz' goodness-of-fit test—have been described in detail to indicate the types of Monte Carlo studies that this text is intended to assist. A brief overview of the subsequent chapters is now given to suggest specifically the potential applications of the material.

One of the first obstacles in conducting a multivariate Monte Carlo study is to obtain or to develop the necessary univariate generation routines. Chapter 2 provides a fairly detailed discussion of the univariate distributions and their generation needed in the sequel. General variate generation techniques are described and illustrated with many examples including the normal and gamma distributions. Two univariate systems of distributions—the Johnson translation system and a generalized exponential power family—are also examined as they can be neatly extended to higher dimensions.

Some multivariate generation preliminaries are provided in Chapter 3. General generation techniques are described and then illustrated on a variety of examples, including the t, Burr, and Bingham distributions.

Chapters 4 through 10 deal with specific multivariate distributions. In Chapter 4 the notation for the multivariate normal is given, and the use of this distribution as a baseline case in Monte Carlo studies is outlined. Probabilistic mixtures of normal variates are also considered. Johnson's translation system is described in Chapter 5. This set of distributions is one of the few that have actually been used extensively in multivariate statistical simulation studies, especially in discriminant analysis work. Elliptically contoured distributions are presented in Chapter 6 with an emphasis on the Pearson Type II and Type VII (including the t and Cauchy distributions) members, since they cover the class fairly well and are easy to use in simulation. Circular and spherical distributions are reviewed in Chapter 7. Although of interest in their own right, the distributions are used in a construction scheme to develop some new distributions that complement the elliptically contoured class. Other departures from the multivariate normal distribution can be found with the distributions based on Khintchine's unimodality theorem (Chapter 8). These relatively new distributions offer some intriguing possibilities. A unifying class for the Burr, Pareto, and logistic distributions is developed in Chapter 9. Until recently, these distributions were thought to be rather distinct. Finally, in Chapter 10 we consider some miscellaneous distributions that have occasionally been used in simulation work. These include the Morgenstern, Plackett, Gumbel, and Ali-Mikhail-Haq distributions. The Wishart distribution is also considered, since many articles have appeared discussing its efficient computer generation.

A significant undercurrent throughout these chapters involves the construction or development of new multivariate distributions for use in Monte Carlo studies. Many "new" distributions are direct multivariate generalizations of certain simple univariate or bivariate distributions. Although this approach leads occasionally to valuable parametric forms, the resulting distributions tend to preserve the same deficiencies or limitations as their lower dimensional counterparts. In a few cases, however, the distributions to be described arise from unique or clever approaches. Chapter 11 is intended to highlight some of these ideas and to present some other possible approaches awaiting further development.

The multivariate distributions presented in this text are intended to be used in future Monte Carlo studies. Moreover, developing multivariate distributions from the standpoint of using them in Monte Carlo work provides a distinct and possibly inspiring perspective. Even relatively simple approaches can lead to useful new distributions (e.g., those in Chapters 7 and 8). Conceivably, other viable multivariate distributions could be invented from this viewpoint. The area of multivariate distributions for Monte Carlo purposes seems fertile for future developments.

CHAPTER 2

Univariate Distributions and their Generation

Methods for generating multivariate distributions rely heavily on techniques and results from the univariate context. This chapter provides a survey of key results in univariate generation for this later use. Aside from the variate generation problem, some specific construction schemes are highlighted in anticipation of subsequent multivariate generalizations. The random variate generation literature is immense (with, some might say, an "import that goes far beyond the dictates of utility," Borwein and Borwein, 1984), so that only a very selective treatment is given here. More comprehensive surveys can be found in Bratley, Fox, and Schrage (1983), Kennedy and Gentle (1980), Knuth (1981), Law and Kelton (1982), and Rubinstein (1981). In this chapter an emphasis is directed toward generation algorithms that are simple but relatively efficient compared to the fastest but typically more complex algorithms.

Section 2.1 provides an introduction to univariate generation through a discussion of general techniques illustrated with examples. The univariate normal distribution has received considerable attention from the generation standpoint. Algorithms that can be extended to higher dimensions are emphasized in Section 2.2. The widely used univariate family known as Johnson's system is considered in Section 2.3. This system has the advantage that it too can be extended to higher dimensions along with the multivariate normal distribution. The generalized exponential power distribution is described in Section 2.4. This distribution is a precursor to the Khintchine distributions of Chapter 8. The gamma distribution that enters into many multivariate constructions is reviewed in Section 2.5. Finally, in Section 2.6 a philosophical discussion of uniform number generation is given, and some suggestions for coping with these generators are revealed.

18

2.1. GENERAL METHODS FOR CONTINUOUS UNIVARIATE GENERATION

Before delving into generation schemes for specific distributions, a review of general methods is given. The taxonomy of methods used here is threefold: inversions of the distribution function, transformations based on special distributional relationships, and acceptance-rejection methods. This delineation is certainly not mutually exclusive since *any* method must involve transforming an independent random source—typically the uniform distribution on the unit interval, denoted uniform 0–1. Attention is first focused on the inverse distribution function approach since it is easy to understand, applicable to many important distributions, and particularly attractive in designing efficient (univariate) Monte Carlo studies. Somewhat more complicated transformations are considered next. Finally, acceptance-rejection methods are described. A departure from traditional expositions is that some of the drawbacks of acceptance-rejection methods are admitted.

Inverse Distribution Function Method

The basic problem common to all three methods of this chapter is to generate a variate X whose distribution is specified by F, a distribution function, or its corresponding density function f. It is assumed that an unlimited source of independent uniform 0–1 variates $U_1, U_2, \ldots$ is available. The extent to which these assumptions (independence and uniformity) are valid in practice is discussed in Section 2.6.

The inverse distribution method is stated simply as:

1. Generate a variate U that is uniform 0–1.
2. Let $X = \sup_x [F(x) \leqslant U]$.

The resulting variate X from step 2 has distribution function F. In the event that F is strictly increasing, then its inverse function F^{-1} is well-defined and step 2 becomes

$$X = F^{-1}(U). \tag{2.1}$$

In almost all cases of practical interest, (2.1) can be applied to accomplish step 2.

The standard illustration of this method is given in Figure 2.1 for the exponential distribution for which $F(x) = 1 - \exp(-x)$. The inverse of F is $F^{-1}(U) = -\ln(1 - U)$ where $0 < U < 1$. Frequently, the streamlined

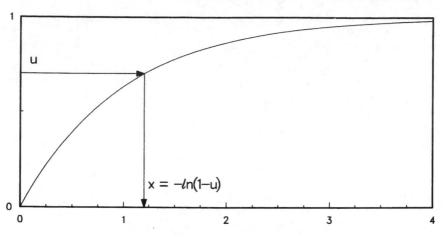

Figure 2.1. Exponential generation.

generation formula $X = -\ln(U)$ is used since U is uniform 0–1 if $1 - U$ is.

As another example, consider the density function given by

$$f(\theta) = \kappa[2\sinh(\kappa)]^{-1}\exp(\kappa\cos\theta)\sin(\theta), \qquad 0 < \theta < \pi, \kappa > 0.$$

This density governs one of the angles of Fisher's distribution on the sphere (Section 7.2). The corresponding distribution function is

$$F(\theta) = \frac{\exp(\kappa) - \exp(\kappa\cos\theta)}{2\sinh(\kappa)}.$$

Solving $F(\theta) = U$ for θ yields

$$\theta = \cos^{-1}\{\kappa^{-1}\ln[(1 - U)\exp(\kappa) + U\exp(-\kappa)]\}.$$

Figure 2.2 depicts this transformation for the case $\kappa = 1$.

The inverse distribution function method has several salient features that warrant discussion. The practical implementation of the method rests with evaluating F^{-1}. This does not imply that a closed-form expression for this function must be available. In particular, an equivalent approach is to solve

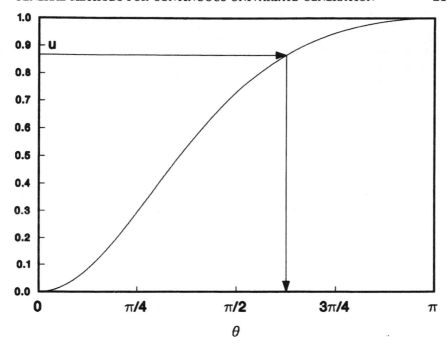

Figure 2.2. Fisher's θ generation.

the equation $F(x) = U$, where U is uniform 0–1. Frequently, the recursion

$$x_{n+1} = x_n - \frac{F(x_n) - U}{f(x_n)} \qquad (2.2)$$

converges in a few iterations to a fixed point x^* such that $F(x^*) = U$. A convenient starting value for this recursion is the median of the distribution. If the median is not known, it can be computed from (2.2) using U equal to 0.5 and x_0 a value in the support of X.

The inverse transformation method is theoretically exact in that if U is really distributed uniform 0–1, then $X = F^{-1}(U)$ has distribution function F. In practice, F^{-1} cannot be evaluated perfectly because of finite computer precision. However, with a little care, six to eight significant digit accuracy is possible with many choices of F. Trouble spots will occur for values of U near 0 or 1 if the support is the whole real line or in general for any region in which F^{-1} is relatively flat. Particular implementations can be checked against tabled percentiles by deliberately supplying as the uniform variate selected p-values such as 0.01, 0.05, 0.1, 0.5, and so forth.

Among distributions that have simple analytical expressions for F^{-1} are the following (including only essential shape parameters):

Burr	$f(x) = ckx^{c-1}[1 + x^c]^{-k-1}$
	$F^{-1}(u) = \left[(1 - u)^{-1/k} - 1\right]^{1/c}$
Cauchy	$f(x) = \left[\pi(1 + x^2)\right]^{-1}$
	$F^{-1}(u) = \tan\left[\pi\left(u - \frac{1}{2}\right)\right]$
Laplace	$f(x) = (\frac{1}{2})\exp| - x|$
	$F^{-1}(u) = -\ln(2u), \qquad u \leqslant \frac{1}{2}$
	$\qquad\quad = -\ln[2(1 - u)], \qquad u > \frac{1}{2}$
Lambda	$f[F^{-1}(u)] = [\lambda_3 u^{\lambda_3 - 1} + \lambda_4 u^{\lambda_4 - 1}]$
	$F^{-1}(u) = u^{\lambda_3} - (1 - u)^{\lambda_4}$
Logistic	$f(x) = [\exp(-x)][1 + \exp(-x)]^{-2}$
	$F^{-1}(u) = -\ln\left(\dfrac{1 - u}{u}\right)$
Pareto	$f(x) = \dfrac{a}{x^{a+1}}$
	$F^{-1}(u) = (1 - u)^{-(1/a)}$
Type I extreme value	$f(x) = \exp(-x)\exp\left[-\exp(-x)\right]$
	$F^{-1}(u) = -\ln\left[-\ln(u)\right]$
Weibull	$f(x) = \beta x^{\beta-1}\exp(-x^\beta), \qquad x > 0$
	$F^{-1}(u) = \left[-\ln(1 - u)\right]^{1/\beta}$

The lambda distribution's density is given in a slightly different form from the others listed above since the distribution itself is defined in terms of its inverse distribution function. To plot its density, $f[F^{-1}(u)]$ is evaluated for values of u in the unit interval and plotted versus $F^{-1}(u)$ (see, e.g., Ramberg, Tadikamalla, Dudewicz and Mykytka, 1979). Some notable omissions from the above list include the normal and gamma distributions, for which we resort to other methods of generation (Sections 2.2 and 2.5).

Perhaps the most important feature of the inverse transformation method in univariate generation is its natural facility for use with variance reduction procedures. Antithetic variates and common random number streams (also known as blocking with the uniforms) are standard devices (Bratley, Fox, and Schrage, 1983) to improve the precision of estimators in simulation studies. A nice review of these and other variance reduction methods is given by Wilson (1984). We do not dwell on this topic here since in the multivariate arena these techniques are much more difficult or even impossible to apply. Some preliminary results in multivariate antithetic sampling are available in Wilson (1983).

Transformation Method

Distributions with intractable inverse distribution functions can sometimes be generated with a transformation other than $F^{-1}(U)$. For example, the t distribution with n degrees of freedom has an inverse distribution function that is difficult to evaluate numerically (Koehler, 1983). The t distribution can be given, however, as

$$t = \frac{X_1}{(X_2/n)^{1/2}},$$

where X_1 is standard normal (Section 2.2) and independent of X_2, which is chi-square with n degrees of freedom (Section 2.5). Several distributions can be generated in a similar fashion, as summarized in Table 2.1.

Aside from simple algorithms, the transformation method is valuable in that it can suggest approaches to constructing multivariate distributions. The multivariate Burr distribution (Chapter 9) can be obtained by a simple extension of the univariate construction used in item 6 of Table 2.1. In particular, the distribution of $Y_i = X_i/X_{n+1}$, $i = 1, 2, \ldots, n$, is multivariate Burr if the X_i's are independent with the first n having standard exponential distributions and with X_{n+1} having a gamma distribution with shape parameter α.

TABLE 2.1.

1. $X \sim N(0,1)$; $Y = X^2$	$\chi^2_{(1)}$
2. $X \sim N(\mu, \sigma^2)$; $Y = \exp(X)$	lognormal
3. $X_1 \sim N(0,1)$; $X_2 \sim \chi^2_{(n)}$; X_1, X_2 independent; $Y = X_1/\sqrt{(X_2/n)}$	$t_{(n)}$
4. $X_1 \sim \chi^2_{(m)}$; $X_2 \sim \chi^2_{(n)}$; X_1, X_2 independent; $Y = (X_1/m)/(X_2/n)$	$F(m, n)$
5. $X_1 \sim \Gamma(\alpha_1, \beta)$; $X_2 \sim \Gamma(\alpha_2, \beta)$; X_1, X_2 independent; $Y = X_1/(X_1 + X_2)$	beta(α_1, α_2)
6. $X_1 \sim \Gamma(1,1)$; $X_2 \sim \Gamma(\alpha, 1)$; X_1, X_2 independent; $Y = X_1/X_2$	Burr(α)

Notation definitions:
  $N(\mu, \sigma^2)$ is a normal distribution with mean μ and variance σ^2.
  $\chi^2_{(n)}$ refers to a chi-squared distribution with n degrees of freedom.
  $\Gamma(\alpha, \beta)$ is a gamma distribution with shape parameter α and scale parameter β.

As a final example, Chambers, Mallows, and Stuck (1976) provide an ingenious transformation for generating stable random variables. In standard form the characteristic function of the stable class is

$$\phi(t; \alpha, \beta) = \exp\left\{-|t|^{\alpha}\exp\left[-\left(\frac{\pi}{2}\right)i\beta\kappa(\alpha)\text{sign}(t)\right]\right\}, \quad 0 < \alpha \leqslant 2, \alpha \neq 1$$

$$= \exp\left\{-|t|\left[1 + \left(\frac{2}{\pi}\right)i\beta \ln|t|\text{sign}(t)\right]\right\}, \quad \alpha = 1,$$

with the skewness parameter β satisfying $-1 \leqslant \beta \leqslant 1$ and $\kappa(\alpha) = 1 - |1 - \alpha|$. In the symmetric case ($\beta = 0$), this class includes the normal ($\alpha = 2$) and Cauchy ($\alpha = 1$) distributions. Taking $W = -\ln U$ where U is uniform 0–1 and Φ an independent uniform on $(-\pi/2, \pi/2)$, the transformation to a stable variate is

$$X = \left[\sin \alpha(\Phi - \Phi_0)\right](\cos \Phi)^{1/\alpha}\left\{W^{-1}\cos[\Phi - \alpha(\Phi - \Phi_0)]\right\}^{(1-\alpha)/\alpha},$$

$$\alpha \neq 1$$

$$= \left(\frac{2}{\pi}\right)\left\{\left[\left(\frac{\pi}{2}\right) + \beta\Phi\right]\tan \Phi - \beta \ln\left(\frac{\pi W \cos \Phi}{\pi + 2\beta\Phi}\right)\right\}, \quad \alpha = 1,$$

where $\Phi_0 = -(\pi/2)\beta\kappa(\alpha)/\alpha$. Chambers and coauthors discuss some numerical aspects of these transformations and provide a FORTRAN program for implementation.

Acceptance-Rejection (AR) Methods

This approach has become very popular in recent years, as evidenced by the increasing number of new algorithms based on AR methods. For example, more than a dozen papers have appeared since 1977 for generating the gamma distribution by AR methods. Other distributions that have attracted interest for this technique include the beta (Schmeiser and Shalaby, 1980), exponential power (Tadikamalla, 1980), and Cauchy (Kronmal and Peterson, 1981). First the AR method is described and then some of its practical aspects are examined.

The basic problem is to generate a variate X having density f as given in Figure 2.3. It is assumed that variates from a related density g can be

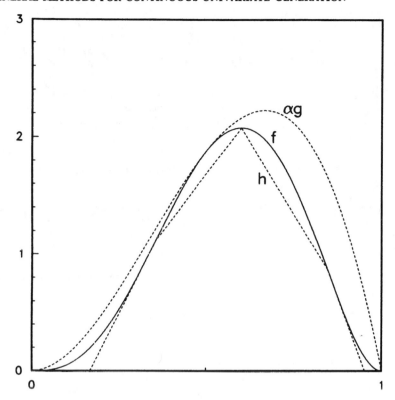

Figure 2.3. Rejection method.

readily generated. The density g approximates f if the value

$$\alpha = \max_x \frac{f(x)}{g(x)}$$

is a constant not too much larger than 1. Of course, if α equals 1, then f is identical to g. From the figure, the function αg is seen to "dominate" f in the sense that $\alpha g(x) \geqslant f(x)$ for all x's in the support of X. The use of generated variates from g to obtain generated variates from f is revealed in the following algorithm, which gives the rejection aspect of the AR method:

1. Generate X from density g and compute $T = f(X)/\alpha g(X)$.

2. Generate U as uniform 0–1.
3. If $U > T$, reject X and return to step 1.
4. If $U \leqslant T$, accept X as a variate from f.

It may happen that f is a time-consuming function to evaluate. If, however, there exists a function h as in the figure for which $h(x) \leqslant f(x)$ for all x in the support, then a "fast" (preliminary) acceptance test can be made. Geometrically, the procedure can be viewed as follows:

1. Generate a point (x, y) that is uniformly distributed in the region bounded by αg and the x-axis.
2. If the point falls below h, accept x immediately.
3. If the point is above h and below f, accept x.
4. Otherwise, reject x and try again at step 1.

As a specific example needed later (Section 7.1), consider the power sine distribution having density function

$$f(\theta; k) = c_k \sin^k \theta, \qquad 0 < \theta < \pi,$$

where k is an integer and

$$c_0 = \frac{1}{\pi}$$

$$c_1 = \tfrac{1}{2}$$

$$c_k = \left(\frac{k}{k-1} \right) c_{k-2}, \qquad k \geqslant 2.$$

A rejection method is devised for the cases $k \geqslant 2$ since for $k = 0$, Θ is uniform on $(0, \pi)$ and for $k = 1$, the expression $\Theta = \cos^{-1}(1 - 2U)$ provides the appropriate variate for U uniform 0–1.

Figure 2.4 illustrates the development of the algorithm. A point (U_1, U_2) is selected having a uniform distribution over the rectangle $(0, \pi) \times (0, 1)$. Note that we can ignore c_k in deriving this algorithm. If the point is in region A, then $X = U_1$ has the required distribution. If the point is in region C_1 or C_2, the corresponding density of $X = U_1$ is proportional to $|\cos^k \theta|$, which is close to what we want. We can in fact use this point by

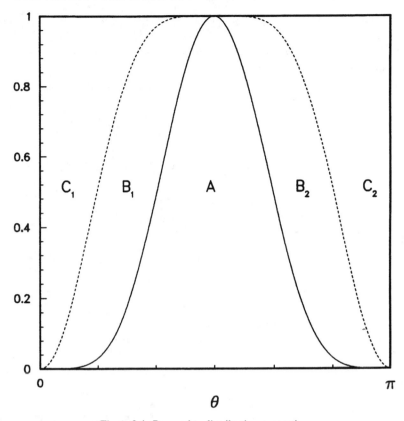

Figure 2.4. Power sine distribution generation.

making a simple adjustment:

$$X = U_1 + \frac{\pi}{2}, \qquad \text{for } 0 < U_1 < \frac{\pi}{2}$$

$$= U_1 - \frac{\pi}{2}, \qquad \text{for } \frac{\pi}{2} < U_1 < \pi.$$

The point (U_1, U_2) is rejected only if it is in regions B_1 or B_2. For $k = 2$, the rejection regions are nil.

To summarize this verbal description, the following algorithm is offered:

1. Generate U_1 and U_2 independent uniform 0–1 variates and set $X = \pi U_1$.
2. (Rejection step.) If $\sin^k X < U_2 < 1 - |\cos^k X|$, return to step 1.
3. If $U_2 \leqslant \sin^k X$, accept $\theta = X$.
4. Otherwise, $U_2 \geqslant 1 - |\cos^k X|$. If $X \leqslant \pi/2$, accept $\theta = X + \pi/2$. If $X > \pi/2$, accept $\theta = X - \pi/2$.

For increasing $k \geqslant 2$, the efficiency declines, as follows:

k	Proportion (U_1, U_2) Leading to Accepted θ Variate
3	0.849
4	0.750
5	0.679
10	0.492
20	0.352
50	0.225

This performance is adequate in many circumstances, especially for use with the distributions to be given in Chapter 7.

AR algorithms are important because they can be very fast in comparison to transformation method algorithms. For some distributions, such as the gamma, they can lead to extremely simple but efficient algorithms as well. Some of the more complicated AR algorithms are also available in commercially available packages such as IMSL (1980). This relieves the occasional user of the onerous burden of correctly implementing a complicated algorithm. Even published FORTRAN versions that are presumably correct must be entered with meticulous care. Thorough scrutiny of output sequences is not likely to detect minor errors.

Another issue that could be raised with AR methods is the importance of fast generation times. In the author's experience, variate generation time typically consumes a small fraction of total computing time in Monte Carlo studies. Had the variate generation time been zero in these investigations, the savings would have gone unnoticed. In many Monte Carlo studies, substantial savings can be accrued either in improving the processing of the generated samples or via a shrewd design.

A final aspect of the AR method is the related "ratio-of-uniforms" technique developed by Kinderman and Monahan (1977). This technique has a number of characteristics in common with AR methods, but the specific formulation with f and the dominating function αg do not exactly apply. The basic idea is as follows: Let $f(x)$ be the density from which variates are to be generated. Define the region $C = \{(U_1, U_2): 0 \leqslant U_1 \leqslant f^{1/2}(U_1/U_2)\}$. If a pair (U_1, U_2) is uniformly distributed over the region C, then $X = U_1/U_2$ has density function f. An application of this method is algorithm GBH, given in Section 2.5, for the gamma distribution.

2.2. NORMAL GENERATORS

A random variable X has a normal distribution denoted $N(\mu, \sigma^2)$ if its density function is

$$\phi(x) = \frac{1}{\sigma\sqrt{2\pi}} \exp\left[-\frac{(x-u)^2}{2\sigma^2}\right], \qquad -\infty < x < \infty \qquad (2.3)$$

with corresponding distribution function

$$\Phi(x) = \int_{-\infty}^{x} \phi(t)\, dt. \qquad (2.4)$$

The standard normal distribution is $N(0,1)$ having zero mean and unit variance. It is sufficient for generation purposes to derive algorithms to generate X as $N(0,1)$, since $Y = \sigma X + \mu$ is $N(\mu, \sigma^2)$.

A popular scheme for generating $N(0,1)$ is due to Box and Muller (1958) and in fact yields a pair of independent $N(0,1)$ variates:

$$X_1 = (-2\ln U_1)^{1/2}\cos(2\pi U_2)$$

$$X_2 = (-2\ln U_1)^{1/2}\sin(2\pi U_2), \qquad (2.5)$$

where U_1 and U_2 are independent uniform 0–1. To see intuitively that this method works, three observations are needed: the point $[\cos(2\pi U_2), \sin(2\pi U_2)]$ is uniformly distributed on the unit circle; $-2\ln U_1$ is $\chi^2_{(2)}$; if X_1 and X_2 are independent $N(0,1)$, then $W = X_1^2 + X_2^2$ is $\chi^2_{(2)}$ and $(X_1/W, X_2/W)$ is uniformly distributed on the unit circle.

An alternative method that avoids the trigonometric evaluations is due to Marsaglia and Bray (1964) and has come to be known as the modified polar method. Pairs of independent uniform 0–1 variates U_1 and U_2 are generated, and the following expressions are evaluated:

$$V_1 = 2U_1 - 1$$

$$V_2 = 2U_2 - 1$$

$$R = V_1^2 + V_2^2.$$

If $R > 1$, then we reject this pair (U_1, U_2) from further consideration and generate new U_i's. For the acceptable U_i's, we then produce the normal variates:

$$X_1 = V_1 \left[\frac{-2\ln(R)}{R} \right]^{1/2}$$

$$X_2 = V_2 \left[\frac{-2\ln(R)}{R} \right]^{1/2}. \tag{2.6}$$

The modified polar method is equivalent, in principle at least, to the Box-Muller transformation. The formulas in (2.6) can be rewritten as

$$X_1 = \frac{U_1}{R^{1/2}} \cdot (-2\ln R)^{1/2}$$

$$X_2 = \frac{U_2}{R^{1/2}} \cdot (-2\ln R)^{1/2}. \tag{2.7}$$

The point (U_1, U_2) restricted by $U_1^2 + U_2^2 \leqslant 1$ has a uniform distribution on the interior of the unit circle. The point $(U_1/R^{1/2}, U_2/R^{1/2})$ is the projection of (U_1, U_2) to the boundary of the unit circle, and thus has a uniform distribution on the circle. Moreover, it can be shown that R is uniform 0–1 so that $-2\ln R$ is $\chi^2_{(2)}$. Hence, both methods involve selecting a point at random on the unit circle and then adjusting its squared distance from the origin to correspond to a $\chi^2_{(2)}$. This specific approach can be generalized to higher dimensions and used with other distance distributions to obtain the elliptically contoured distributions in Chapter 6.

Some extremely fast normal generators exist including the so-called "super-duper" generator based on a paper by Marsaglia, MacLaren, and Bray (1964). More recently, Kinderman and Ramage (1976) proposed a fast but reasonably simple AR scheme. These generators are not necessary for multivariate generation or construction, but if already residing on the computer system can be employed.

2.3. JOHNSON'S TRANSLATION SYSTEM

In applying the transformation method for generating variates, the specific functional forms used to generate these distributions were merely stated. No guidance was provided to address the problem of finding a transformation in cases in which it is not already "well-known." Another approach in using the transformation method is to start with a useful, easy-to-generate distribution such as a $N(\mu, \sigma^2)$ and consider the following four transformations:

$$Y = X \qquad\qquad\qquad \text{Normal}$$
$$Y = \lambda \exp(X) + \xi \qquad\qquad \text{Lognormal}$$
$$Y = \lambda[1 + \exp(X)]^{-1} + \xi \qquad \text{Logit-normal}$$
$$Y = \lambda \sinh(X) + \xi \qquad\qquad \text{Sinh}^{-1}\text{-normal} \qquad (2.8)$$

where $\sinh(X) = [\exp(X) - \exp(-X)]/2$. These forms constitute the sim-

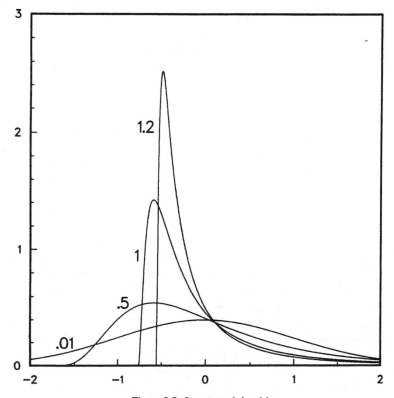

Figure 2.5. Lognormal densities.

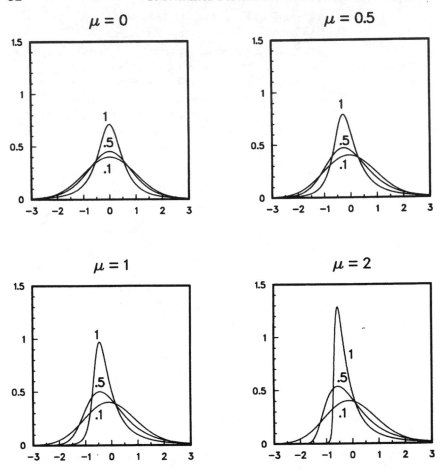

Figure 2.6. Sinh^{-1}-normal densities.

ple formulas needed for variate generation. The parameters λ and ξ are introduced to control the scale and location of Y but do not affect its shape. The support of Y varies with the specific transformation. For the normal and sinh^{-1}-normal distributions, the support is $(-\infty, \infty)$. For the lognormal the support is (ξ, ∞) and for the logit-normal the support is $(\xi, \xi + \lambda)$. The shapes of these distributions can be quite different as evidenced in Figures 2.5–2.7.

Plots of lognormal densities appear in Figure 2.5. The parameter μ was taken to be zero since it acts as a scale parameter in this transformed normal distribution and scale can be more conveniently controlled by λ. The labeled curves correspond to four values (0.01, 0.5, 1, 1.2) of σ, which

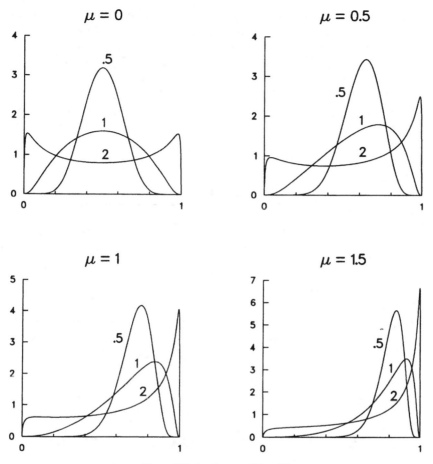

Figure 2.7. Logit-normal densities.

clearly can be seen to be a shape parameter. The parameters λ and ξ were determined so that the lognormal distributions depicted have zero mean and unit variance. This results in distributions with different supports. Similar decisions were made for the $\sinh^{-1}$-normal densities displayed in Figure 2.6. A dozen combinations of the parameters $\mu(0, 0.5, 1, 2)$ and σ (0.1, 0.5, 1) were considered, with λ and ξ calculated to yield zero mean and unit variance. Additional flexibility is obtained with the $\sinh^{-1}$-normal distribution over the lognormal. As μ gets large, the $\sinh^{-1}$-normal distribution tends to a lognormal.

Density plots for the logit-normal distribution appear in Figure 2.7. No attempt was made here to control the mean and variance of this transformed normal distribution (i.e., $\lambda = 1$, $\xi = 0$) since the necessary formulas are very messy (Johnson, 1949b) and for plotting purposes the support is $[0, 1]$. As in the $\sinh^{-1}$-normal case, a dozen parameter combinations of (μ, σ) are given that indicate rather diverse shapes obtainable over the finite support $(0, 1)$. Skewed, symmetric, bimodal, unimodal, and/or antimodal distributions can be obtained with proper choice of the parameters.

The Johnson translation system continues to attract the interest of statistical researchers (Slifker and Shapiro, 1980 and Mage, 1980). The primary reason for its popularity stems in applications from the strategy of applying the inverses of the transformations in (2.8) to *data* in hopes of subsequently invoking normal distribution theory analysis. This was the tactic that evidently motivated the development of the system originally and that continues to provide a compelling reason for its use today.

Chapter 5 provides an extension of the Johnson system to a multivariate setting. For many Monte Carlo studies, it is helpful to be able to specify the means, variances, and covariances in the non-normal multivariate distributions. Results for this purpose are given.

A final comment on the univariate system is that the transformations given in (2.8) can be reasonably applied to any univariate distribution. Of course, the degree to which the resulting distributions are valuable depend on the distribution of X. Tadikamalla and Johnson (1982a, b) have developed a system of distributions using the logistic distribution as a starting point. This system could be extended to higher dimensions using the material in Chapter 5 as a guide.

2.4. GENERALIZED EXPONENTIAL POWER DISTRIBUTION

In the Johnson translation system of the previous section, only the normal distribution member is symmetric. In many Monte Carlo robustness or power studies, symmetric alternatives to the normal distribution may be of interest. In this section, a distribution useful for this purpose is described (Johnson, Tietjen, and Beckman, 1980). The distribution is also included since it inspires extensions to several multivariate distributions in Chapter 8.

This distribution arises from the following considerations. Let X be uniformly distributed on the interval $(-R, R)$ where R itself is a random variable distributed as $\sqrt{\chi^2_{(3)}}$. Upon integrating out the distribution of R, the random variable X has unconditionally a standard normal distribution $N(0, 1)$. Two obvious generalizations in this construction involve using a gamma variate with shape parameter α and an arbitrary power, say τ, to which the gamma variate is raised. Scale (σ) and location (μ) parameters can also be introduced to yield the following density function:

$$f(x) = \frac{A}{2\sigma\Gamma(\alpha)} \int_{\left[\frac{A|x-\mu|}{\sigma}\right]^{1/\tau}}^{\infty} w^{\alpha-\tau-1} e^{-w}\, dw, \qquad (2.9)$$

where $-\infty < x, \mu < \infty$; $\alpha, \tau, \sigma > 0$; $A = [\Gamma(\alpha + 2\tau)/3\Gamma(\alpha)]^{1/2}$.

This functional form can be viewed as the tail area of a gamma density with shape parameter $\alpha - \tau$. The parameters in (2.9) are readily described: μ—location; σ—scale; α—shape (one-half of the degrees of freedom); τ—shape (the power to which the gamma variate is raised). The particular parameterization given in (2.9) was devised so that the mean of X is μ and its variance is σ^2. The density f is symmetric and unimodal at μ. The kth moment of X for $\mu = 0$ and $\sigma^2 = 1$ is

$$E(X^k) = \begin{cases} 0 & k \text{ is odd} \\[2mm] \dfrac{\Gamma(\alpha + kt)}{(k+1)\Gamma(\alpha)A^k} & k \text{ is even.} \end{cases}$$

Thus the coefficient of kurtosis is easily determined as

$$\beta_2 = \frac{9\Gamma(\alpha + 4\tau)\Gamma(\alpha)}{5\Gamma^2(\alpha + 2\tau)}.$$

Since there are two parameters, α and τ, in this expression, it should be possible to identify many pairs that yield the same kurtosis. Figure 2.8 provides some densities that correspond to distributions with zero mean, unit variance, and common kurtosis as indicated. The diversity of shapes near the mode may be surprising. The tails of the distributions, however, must become coincident for the kurtosis values to be the same.

Besides the normal distribution ($\alpha = 1.5$, $\tau = 0.5$), several other special distributions can be identified in (2.9). In particular, if $\alpha - \tau = 1$, the integral simplifies to a form recognized as the exponential power distribution (Subbotin, 1923 and Box and Tiao, 1973). Special cases of this distribution are the uniform ($\alpha = 1$, $\tau \to 0$), the normal, and the Laplace or double exponential ($\alpha = 2$, $\tau = 1$) distributions. Figure 2.9 presents for a variety of (α, τ) values some density plots of (2.9).

The construction of this distribution makes it well-suited for use in Monte Carlo studies. The following algorithm generates variates according

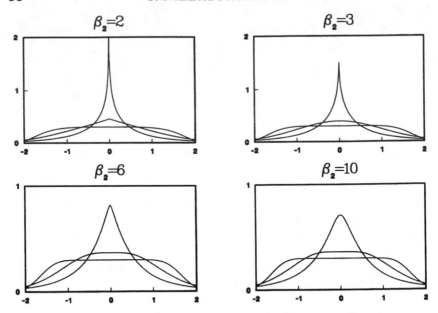

Figure 2.8. Generalized exponential power densities, common kurtosis.

to density (2.9).

1. Generate W having a gamma distribution with shape parameter α and scale parameter 1.
2. Set $Y = W^\tau$.
3. Generate V having a uniform 0–1 distribution.
4. Set $U = 2V - 1$.
5. Set $X = \sigma[3\Gamma(\alpha)/\Gamma(\alpha + 2\tau)]^{1/2}YU + \mu$.

Algorithms GS or GBH from Section 2.5 can be used in step 1 depending on the value of α.

The structure of the generation algorithm lends itself nicely to a variance reduction design if the effects due to the choice of τ are of interest. For a fixed value of α, the same set of generated gamma variates can be used for various choices of τ, thus providing correlated samples. Reuse of gamma variates as mentioned here is analogous to the blocking on uniforms described in Section 2.1. Because of potential gains due to this design, the

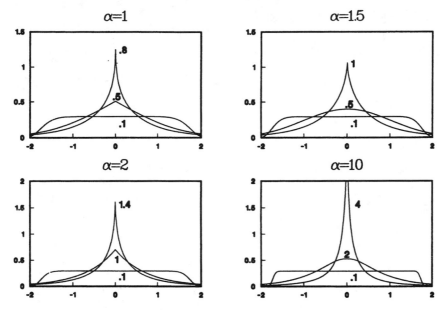

Figure 2.9. Generalized exponential power densities, common α.

algorithm given here should probably be used in lieu of algorithms appropriate only for the exponential power distribution special case (Johnson, 1979 and Tadikamalla, 1980). Two recent simulation studies that employed generated variates from (2.9) include those of Bartels (1982) and Beauchamp and Kane (1982).

Aside from intrinsic interest, this distribution provides a useful starting point for constructing new multivariate families of distributions. The construction scheme can be viewed equivalently as $X = ZU$, where U is uniform 0–1, and independent of Z, which has a reflected gamma to a power distribution. In n dimensions, consider

$$X_i = Z_i U_i \qquad i = 1, 2, \ldots, n,$$

where for each i, Z_i and U_i are independent and U_i is uniform 0–1. The distribution of Z_i is selected to yield a specified distribution for X_i. The set of Z_i's need not be independent nor does the set of U_i's need to be independent. This flexibility in dependence is explored in Chapter 8 for X_i's that have normal or exponential distributions.

2.5. GAMMA GENERATORS

The gamma distribution is denoted $\Gamma(\alpha, \beta)$ and has density function

$$f(x) = \frac{1}{\Gamma(\alpha)\beta^{\alpha}} x^{\alpha-1} \exp\left(-\frac{x}{\beta}\right), \qquad x > 0, \alpha, \beta > 0. \qquad (2.7)$$

If X is $\Gamma(\alpha, 1)$, then βX is $\Gamma(\alpha, \beta)$, so that for generation purposes, it is sufficient to consider $\Gamma(\alpha, 1)$. A plot of selected gamma densities appears in Figure 2.10.

Tadikamalla and Johnson (1981) have surveyed many of the gamma generation algorithms. The algorithms are naturally separated into two cases: $\alpha \geqslant 1$ and $\alpha < 1$, which correspond respectively to finite and infinite values of the gamma density at its mode. The mode occurs at $x = 0$ for $\alpha \leqslant 1$ and at $x = \alpha - 1$ for $\alpha \geqslant 1$.

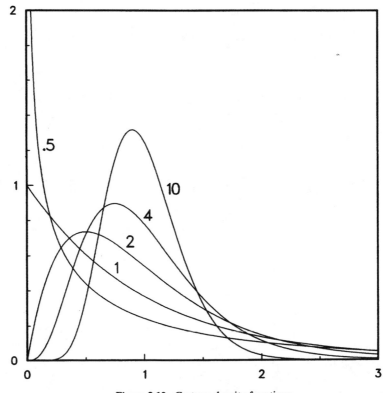

Figure 2.10. Gamma density functions.

Of the six algorithms surveyed by Tadikamalla and Johnson for $\alpha < 1$, the algorithm denoted GS due to Ahrens and Dieter (1974) can be recommended. The algorithm requires only 14 lines of FORTRAN code, applies for all values of α in $(0, 1)$, and is reasonably efficient. Algorithm GS is based on the AR method with the density g given by

$$
g(x) =
\begin{cases}
\dfrac{e\alpha}{e + \alpha} x^{\alpha-1}, & 0 < x \leqslant 1 \\[4mm]
\dfrac{e\alpha}{e + \alpha} e^{-x}, & x > 1.
\end{cases}
$$

It is easy to show that $[(e + \alpha)/e\alpha\Gamma(\alpha)]g(x) \geqslant f(x)$ for all $x > 0$ and

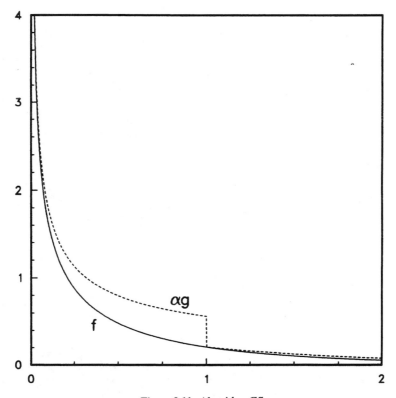

Figure 2.11. Algorithm GS.

$0 < \alpha < 1$. This density g is a mixture of a simple beta distribution for the interval $(0, 1]$ and an exponential distribution conditioned on $X > 1$. Both of these two distributions in the mixture can be generated by inverse transformation. Ahrens and Dieter show that the expected number of uniform 0–1 variates is at most 2.78 if $\alpha = 0.8$ or has an expected value of 2.54 assuming α is uniform 0–1. The functions used in algorithm GS are illustrated in Figure 2.11 for $\alpha = \frac{1}{2}$.

For $\alpha > 1$, algorithm GBH developed by Cheng and Feast (1979) seems a reasonable choice among the many gamma algorithms published recently. Algorithm GBH is based on the ratio-of-uniforms technique described in Section 2.1 in the context of AR methods. A region C is defined by

$$C = \left\{ (U_1, U_2) : 0 \leqslant U_1 \leqslant g\left(\frac{U_1}{U_2} \right) \right\},$$

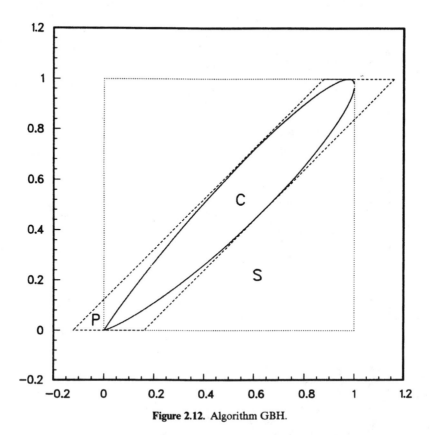

Figure 2.12. Algorithm GBH.

where $g(x) = [x^{4\alpha-1}e^{-x^4}]^{1/2}$. Selecting a point uniformly over C is accomplished by enclosing it within a square S and a parallelogram P, as illustrated in Figure 2.12 for $\alpha = 2.5$. Cheng and Feast recommend the order of operations: Sample a point with the uniform distribution over P; if the point lies in S then check if it also lies in C.

Algorithm GBH provides a reasonable compromise to the usually conflicting criteria of efficiency with respect to execution time and complexity in terms of length of computer code. The version of GBH given by Tadikamalla and Johnson requires only 22 lines of FORTRAN statements. This algorithm is valid for all $\alpha > \frac{1}{4}$ and is in fact, faster in execution than GS over this range. If no shape parameter values less than or equal to $\frac{1}{4}$ are anticipated, then algorithm GBH will suffice.

2.6. UNIFORM 0–1 GENERATORS

A fundamental and sometimes vexing problem in conducting Monte Carlo studies is to draw samples from the uniform distribution on the unit interval $(0, 1)$. Subsequent transformations can convert these uniforms to nearly any distribution of practical interest. The "quality" of the transformed values is therefore dependent on the quality of the generated uniforms: Are they really distributed with density function

$$f(x) = 1, \qquad 0 < x < 1$$

and are multiple realizations independent? If the answers to these two questions are both affirmative, then all of the algorithms given in this text indeed generate samples from the specified distributions.

There is a vast literature devoted to uniform generators, and the intent in this section is certainly not to review all of this material. Relevant references include Knuth (1981), Kennedy and Gentle (1980), and Dudewicz and Ralley (1981). Rather, the point here is to describe briefly the uniform 0–1 generator currently recommended by the author for use in Monte Carlo studies. The recommended generator is Lewis and Payne's (1973) generalized feedback shift register (GFSR) generator. Any such recommendation is couched with numerous caveats including the following two:

1. At least a couple of days effort is required to install the generator on a system if it does not already reside there. Lewis and Payne provide adequate documentation to perform the job, but handling specific computer installation nuances can be time-consuming.

2. If a better generator becomes available, then loyalty to GFSR will quickly evaporate. This is especially true if a deficiency is found with

GFSR. Ramshaw and Amer (1982) claim some difficulties in GFSR that the author and a colleague have been unable to confirm.

The GFSR generator is based on the theory of primitive trinomials. It is a Tausworthe-type generator that can shift bits by means of "exclusive or" operations. Specific details of the generator are omitted here, since Lewis and Payne give a readable account of the underlying theory and implementation procedure. The key properties of GFSR that motivate its use are:

1. The period is not dependent on the computer word size. A period of $2^{98} - 1$ is obtained for the generator based on $x^{98} + x^{27} + 1$.
2. Better n-space uniformity is achieved by GFSR than by Lehmer congruential multiplicative generators. This property refers to a generator's capability to provide points throughout an n-dimensional unit hypercube (Fushimi and Tezuka, 1983).
3. To date, GFSR has passed a large battery of statistical tests designed to identify subtle differences from uniformity or independence.
4. The same basic algorithm works on any machine. Lewis and Payne provide a standard FORTRAN subprogram for use in implementation and generation of uniform 0–1 variates.

Throughout the remainder of this text, the assumption is made that the uniform 0–1 generator does provide independent uniform 0–1 variates. Minor deviations from this assumption will probably be insufficient to invalidate the conclusions of most Monte Carlo studies.

CHAPTER 3

Multivariate Generation Techniques

With the exception of multivariate distributions having independent components, variate generation is much more involved with multivariate distributions than with univariate distributions. The obvious difference is that in the generation of multivariate distributions, the dependencies among the components of the random vector must be handled, which is frequently a nontrivial task. Three general-purpose methods for multivariate generation are described in this chapter. These methods are conditional distribution, transformation, and rejection. As illustration, algorithms are developed for the multivariate Cauchy and Burr distributions and for Bingham's (1974) distribution on the sphere.

3.1. CONDITIONAL DISTRIBUTION APPROACH

The beauty of this conditional distribution approach is that it reduces the problem of generating a p-dimensional random vector into a series of p univariate generation problems. This strategy can utilize the vast body of techniques available for univariate generation. Let the random vector $\mathbf{X} = (X_1, X_2, \ldots, X_p)'$ be the p-dimensional random vector of interest. The conditional distribution approach involves the following steps:

1. Generate $X_1 = x_1$ from the marginal distribution of X_1.
2. Generate $X_2 = x_2$ from the conditional distribution of X_2 given $X_1 = x_1$.
3. Generate $X_3 = x_3$ from the conditional distribution of X_3 given $X_1 = x_1$ and $X_2 = x_2$.
   $$\vdots$$

and so forth through the p steps. Implementation for a specific choice of $\mathbf{X}$ requires the determination of p univariate distributions and their appropriate generation scheme. In many cases of practical interest, the conditional distributions are recognizable although their particular parameters may be fairly complex functions of the variates generated at previous steps. The method may be ascribed to Rosenblatt (1952), who suggested further that the steps of the algorithm be accomplished by the inverse transformation method at each step.

As an illustration, consider the multivariate Cauchy distribution (Johnson and Kotz, 1972, p. 294) having density function

$$f(\mathbf{x}) = \pi^{-(p+1)/2}\Gamma\left(\frac{p+1}{2}\right)(1 + \mathbf{x}'\mathbf{x})^{-(p+1)/2}, \qquad \mathbf{x} \in R^p.$$

The pertinent distributional results are that each component X_i of $\mathbf{X}$ has a univariate Cauchy distribution and that the conditional distribution of

$$\left[m^{1/2}\left(1 + \sum_{i=1}^{m-1} X_i^2\right)^{-1/2}\right]X_m,$$

given $X_1 = x_1, \ldots, X_{m-1} = x_{m-1}$, is univariate Student's t with m degrees of freedom. The Cauchy variate can be generated as $X_1 = \tan[\pi(U - \frac{1}{2})]$, where U is uniform 0–1. The t variates are readily sampled as $Y/\sqrt{(Z/m)}$, where Y is standard normal (Section 2.2) and Z is an independent $\Gamma(m/2, 2)$ variate (Section 2.5).

As another example, suppose $\mathbf{X}$ has a multivariate Burr distribution with density function

$$f(\mathbf{x}) = \left[\frac{\Gamma(k+p)}{\Gamma(k)}\right]\left(1 + \sum_{i=1}^{p} x_i^{c_i}\right)^{-(k+p)} \prod_{i=1}^{p}(c_i x_i^{c_i-1}).$$

Here again the marginal and conditional distributions have particularly convenient (from the variate generation standpoint) forms. In particular, the marginal distribution functions are

$$F(x_i) = 1 - (1 + x_i^{c_i})^{-k},$$

which correspond to Burr's (1942) univariate distribution. Moreover, the conditional distributions are themselves scaled Burr variates, so that an algorithm could be constructed (for the details, see Johnson and Kotz, 1972, pp. 288–290). An explicit method for the Burr distribution is given below.

Subsequent applications of the conditional distribution method include algorithms developed for the Morgenstern (Section 10.1) and Plackett (Section 10.2) bivariate distributions.

3.2. TRANSFORMATION APPROACH

If the conditional distributions of **X** are difficult, then perhaps a more convenient transformation can be found. The tack is to represent **X** as a function of other, usually independent, univariate random variables, each of which can be easily generated. An example of this type of approach was the Box-Muller transformation (Section 2.2), which uses two independent uniform 0–1 variates and converts them to two independent normal variates. Also, stable variates with characteristic exponent α and skewness parameter β (Section 2.1) can be generated by a transformation of independent exponential (W) and uniform (Φ) variates. As a final example from Chapter 2, the generalized exponential power distribution can be obtained as the product of independent uniform and gamma (to a power) variates.

As illustration in the multivariate setting, the multivariate Cauchy and Burr distributions are again used. For the multivariate Cauchy, we can take

$$X_i = \frac{Z_i}{W},$$

where the Z_i's and W are independent, the Z_i's are standard normal, and W is the square root of a $\Gamma(\frac{1}{2}, 2)$ variate.

For the multivariate Burr, we use the transformation

$$X_i = \frac{Y_i}{Z}, \qquad i = 1, 2, \ldots, n,$$

where $Y_1, \ldots, Y_n$, and Z are independent, the Y_i's are standard exponential, and Z is gamma with shape parameter α. More information on the multivariate Burr distribution is provided in Chapter 9.

Other forthcoming applications of the transformation method include the multivariate normal distribution (Section 4.1), which can be generated by a linear transformation of independent univariate normal variates. The Johnson translation system (Chapter 5) arises by application of the exponential, hyperbolic sine, or inverse logit transformations to the components of a multivariate normal distribution. Similarly, many of the elliptically contoured distributions in Chapter 6 emerge naturally from relationships to the multivariate normal distribution. The new multivariate forms in Chapter 7 are based on extensions of transformations used in Chapter 6.

The above examples illustrate the point that the transformation method has wide applicability. However, by virtue of its general nature, it is not always apparent how to find a particular transformation to generate a multivariate distribution X, specified perhaps by its density function f. The following guidelines are given to assist in the search.

1. Starting with the functional form f, apply invertible transformations to the components of X in hopes of obtaining a recognizable distribution. The probability integral transform ($F(X_i)$ where F_i is the distribution function of X_i) could be tried and the results compared to known multivariate distributions with uniform marginal distributions. Incidently, this was the exercise that revealed that the multivariate Burr, Pareto, and logistic distributions are intrinsically related (Chapter 9).

2. Consider transformations of X that simplify arguments of transcendental functions in the density f. For example, if one factor of f is $\exp(-x_1^3 - x_2^3)$, consider the transformation $y_i = x_i^3$. More generally, one might try other component transformations, such as $1/(1 + x)$, $\ln x$, or $\exp(-x)$, that can alter the support of the components to $(0, 1)$, $(0, \infty)$, or $(-\infty, \infty)$. Before embarking on this transformation expedition, it would be wise to set location and scale parameters to nominal (invisible) values. For x_i appearing as $(x_i - \psi)/\lambda$, take $\psi = 0$ and $\lambda = 1$.

3. Attempt to decompose f as the (probabilistic) mixture

$$f(\mathbf{x}) = pf_1(\mathbf{x}) + (1 - p)f_2(\mathbf{x}),$$

where f_1 and f_2 are recognizable and presumably easy to generate. This strategy is used in Section 10.3 to generate one of the conditional distributions in Gumbel's exponential distribution.

4. Check if $f(\mathbf{x})$ can be written as $f(p(\mathbf{x}))$ where $p(\mathbf{x})$ is a quadratic form in x. A generation method for f is immediate from results given in Chapter 6.

5. Another useful tactic is to track down earlier references to the distribution. Possibly an earlier author or the inventor of the distribution was motivated by a physical process that could be modeled via mixtures, convolutions, or products of random variables.

3.3. REJECTION APPROACH

The AR methods that have found such extensive use in univariate generation have not had much impact in multivariate generation. Partly this is due to the relatively limited attention researchers have paid to multivariate generation but more appropriately to some significant practical difficulties. The theory developed in Section 2.1 carries over directly to higher dimensions, so validity is not the problem. The difficulty is in finding a dominating function αg for f if the dependence among the components of $\mathbf{X}$ is strong. A logical choice for g is the density corresponding to independent components with the same marginal distributions as $\mathbf{X}$. In most cases, however, as the dependencies in $\mathbf{X}$ increase, the extent to which g approximates f decreases, and thus the efficiency approaches zero. More complicated choices of g have the disadvantage of making the search for $\alpha = \sup f(\mathbf{x})/g(\mathbf{x})$ where $\mathbf{x} \in R^p$ more difficult.

One idea for negotiating these difficulties is to consider rejection in conjunction with a decomposition of f. In particular, try to write f as

$$f(\mathbf{x}) = pf_1(\mathbf{x}) + (1 - p)[f(\mathbf{x}) - pf_1(\mathbf{x})],$$

where $0 < p < 1$, $f(\mathbf{x}) - pf_1(\mathbf{x}) \geq 0$ for all $\mathbf{x}$, and f_1 is a "similar" density to f but is easy to generate. If p is close to 1, then most of the time f can be generated as f_1. The rest of the time samples from $f(\mathbf{x}) - pf_1(\mathbf{x})$ are needed and perhaps a rejection technique could be used on this distribution. In this framework, inefficiencies in sampling from $f - pf_1$ could be offset somewhat by the gains in using f_1.

As a simple illustration of multivariate rejection, consider Bingham's (1974) distribution on the sphere with density function

$$g(\theta, \phi; \kappa) = [4\pi d(\kappa)]^{-1} \left\{ \exp\left[(\kappa_1 \cos^2 \phi + \kappa_2 \sin^2 \phi) \sin^2 \theta \right] \sin \theta \right\}$$

$$\equiv ce(\theta, \phi), \qquad 0 < \theta < \pi, 0 < \phi < 2\pi,$$

where the proportionality constant $d(\kappa) = {}_1F_1(\frac{1}{2}; \frac{3}{2}; \kappa)$, the confluent hypergeometric function with matrix argument κ (Herz, 1955). Note that the two angles ϕ and θ determine a point on the surface of the sphere.

We could develop a generation scheme using simply $\sin \theta$ as the dominating function, but this is dreadfully inefficient ($\sin \theta$ does not match g very closely). Rather we apply Atkinson's (1982) bipartite rejection scheme, which uses two dominating functions, depending on the value of θ. First, recognize that $g(\pi/2 - \theta, \phi) = g(\pi/2 + \theta, \phi)$, so that g is symmetric about $\theta = \pi/2$, and hence, we may restrict attention initially to $0 < \theta < \pi/2$. Splitting the interval $(0, \pi/2)$ into the two intervals $(0, \pi/3]$ and $(\pi/3, \pi/2)$ motivates the functions satisfying

$$e(\theta, \phi) \leq \exp(k \sin^2 \theta) \sin(2\theta) \equiv h(\theta, \phi), \qquad 0 < \theta \leq \pi/3$$

$$\leq \exp(k) \sin \theta, \qquad\qquad\qquad \pi/3 < \theta < \pi/2,$$

where $k = \max(k_1, k_2)$. We use $\sin \theta$ over the region $(\pi/3, \pi/2)$ where it does resemble g. The function $h(\theta, \phi)$ approximates g and Θ can be generated by inverting its distribution function. Since ϕ does not enter explicitly in either dominating function, we recognize its distribution to be uniform on $(0, 2\pi)$ and independent of Θ. Hence, a variate (Θ, Φ) with density function proportional to $h(\theta, \phi)$ is generated as

$$\Phi = 2\pi U$$

$$\Theta = \sin^{-1} \sqrt{k^{-1}\ln[V(\exp(3k/4) - 1)]} \, ,$$

where U and V are independent uniform 0–1. Atkinson's bipartite rejection scheme requires the calculation of the following quantities:

$$\Delta_1 = \int_0^{2\pi} \int_0^{\pi/3} h(\theta, \phi) \, d\theta \, d\phi$$

$$= \frac{2\pi[\exp(3k/4) - 1]}{k}$$

$$\Delta_2 = \int_0^{2\pi} \int_0^{\pi/3} \sin \theta \, d\theta \, d\phi$$

$$= \pi$$

$$S_1 = \sup_{\theta, \phi} \frac{e(\theta, \phi)}{h(\theta, \phi)}$$

$$= 1$$

$$S_2 = \sup_{\theta, \phi} \frac{e(\theta, \phi)}{\sin(\theta)}$$

$$= \exp(k).$$

With probability $(\Delta_1/S_1)/(\Delta_1/S_1 + \Delta_2/S_2)$, a variate (θ^*, ϕ^*) is sampled from $h(\theta, \phi)$ and is accepted with probability $e(\theta^*, \phi^*)/h(\theta^*, \phi^*)$. Similarly, with probability $(\Delta_2/S_2)/(\Delta_1/S_1 + \Delta_2/S_2)$, a variate (θ^*, ϕ^*) is sampled from $\sin(\theta)$ and is accepted with probability $e(\theta^*, \phi^*)/[\exp(k)\sin(\theta^*)]$. The beauty of Atkinson's scheme is that the proportionality constant $d(k)$ does not enter into the calculations at all, and that the actual construction of the algorithm reduces to some straightforward derivations.

The three schemes outlined in this chapter get considerable play in the remainder of this text. The presentation is intended to be sufficiently general to permit the application of these methods to multivariate distributions not specifically mentioned in this book.

CHAPTER 4

Multivariate Normal and Related Distributions

The multivariate normal distribution is the first and sometimes only multivariate distribution encountered by students in their formal statistical training. This phenomenon is partly due to the pervasive impact of the central limit theorem and partly by default. There are simply not many suitable alternatives to the multivariate normal distribution. Section 4.1 surveys those basic properties of the multivariate normal distribution that serve as a basis by which other continuous multivariate distributions can be compared. Variate generation schemes are also provided in this section. Section 4.2 covers probabilistic mixtures of normal variates. The distributions are natural extensions of the contaminated normal model studied extensively in the robustness-outlier literature (Beckman and Cook, 1983). The plots for the bivariate case ought to provide a good indication of the range of possible distributions.

4.1. MULTIVARIATE NORMAL DISTRIBUTION

A p-dimensional random vector $\mathbf{X} = (X_1, \ldots, X_p)'$ is defined to have the multivariate normal distribution if and only if every nontrivial linear combination of the p-components of $\mathbf{X}$ has a univariate normal distribution. The distribution of $\mathbf{X}$ is denoted $N_p(\mu, \Sigma)$, where μ is a $p \times 1$ mean vector with entries $\mu_i = E(X_i)$ and Σ is a $p \times p$ covariance matrix whose (i, j)th entry is $\text{Cov}(X_i, X_j)$. If the matrix Σ is singular, then with probability one the distribution of $\mathbf{X}$ is confined to a subspace in R^p. If the matrix

49

Σ has full rank p, then the density function of $\mathbf{X}$ exists as

$$f(\mathbf{x}) = (2\pi)^{-p/2}|\Sigma|^{-1/2}\exp\left[-\tfrac{1}{2}(\mathbf{x} - \mu)'\Sigma^{-1}(\mathbf{x} - \mu)\right], \qquad (4.1)$$

with support R^p. The distribution of $\mathbf{X}$ can be represented as a linear transformation of p independent normal variates in $\mathbf{Y} = (Y_1, Y_2, \ldots, Y_p)'$,

$$\mathbf{X} = A\mathbf{Y} + \mu,$$

with A any $p \times p$ matrix for which $AA' = \Sigma$. For many Monte Carlo applications, the full rank case corresponding to the density in (4.1) should suffice. Unless otherwise indicated, the properties presented below are true in general for the $N_p(\mu, \Sigma)$ distribution.

Marginal Distributions

All marginal distributions of $\mathbf{X} \sim N_p(\mu, \Sigma)$ are normal. Explicitly, let $\mathbf{X}$ and the following arrays be partitioned as

$$\mathbf{X} = \begin{bmatrix} \mathbf{X}_1 \\ \mathbf{X}_2 \end{bmatrix}, \qquad \mu = \begin{bmatrix} \mu_1 \\ \mu_2 \end{bmatrix}$$

$$\Sigma = \begin{bmatrix} \Sigma_{11} & \Sigma_{12} \\ \Sigma_{21} & \Sigma_{22} \end{bmatrix}$$

$$\text{with dimensions} \quad \begin{bmatrix} k \times k & k \times (p - k) \\ (p - k) \times k & (p - k) \times (p - k) \end{bmatrix}.$$

The distribution of $\mathbf{X}_1$ is $N_k(\mu_1, \Sigma_{11})$ and that of $\mathbf{X}_2$ is $N_{p-k}(\mu_2, \Sigma_{22})$.

Conditional Distributions

The conditional distribution of $\mathbf{X}_1$ given $\mathbf{X}_2 = \mathbf{x}_2$ is

$$N_k\left[\mu_1 + \Sigma_{12}\Sigma_{22}^{-1}(\mathbf{x}_2 - \mu_2), \Sigma_{11} - \Sigma_{12}\Sigma_{22}^{-1}\Sigma_{21}\right],$$

provided Σ_{22} is nonsingular. Thus the conditional mean $E(\mathbf{X}_1|\mathbf{X}_2 = \mathbf{x}_2)$ is a linear function of $\mathbf{x}_2$, and the covariance of the conditional distribution does not depend on $\mathbf{x}_2$.

Linear and Quadratic Forms

Let $\mathbf{X}$ be $N_p(\mu, \Sigma)$ and let A be $m \times p$ and $\mathbf{a}$ be $m \times 1$ arrays. The distribution of $A\mathbf{X} + \mathbf{a}$ is

$$N_m(A\mu + \mathbf{a}, A\Sigma A').$$

If Σ is nonsingular, the $\mathbf{Y} = \Sigma^{-1/2}(\mathbf{X} - \mu)$ is $N_p(\mathbf{0}, I)$ where $\mathbf{0}$ is the $p \times 1$ zero vector, I is the p-dimensional identity matrix, and $\Sigma^{-1/2}$ is a matrix for which $(\Sigma^{-1/2})(\Sigma^{-1/2}) = \Sigma^{-1}$. In this case $\mathbf{Y}'\mathbf{Y}$ has a $\chi^2_{(p)}$ distribution. In terms of the original $\mathbf{X}$ vector, we have

$$Z = (\mathbf{X} - \mu)'\Sigma^{-1}(\mathbf{X} - \mu)$$

is $\chi^2_{(p)}$, if Σ is nonsingular. This result can be verified upon noting that the density in (4.1) is a function of $\mathbf{x}$ only through $(\mathbf{x} - \mu)'\Sigma^{-1}(\mathbf{x} - \mu)$. In general, densities of this type belong to the elliptically contoured class described in Chapter 6.

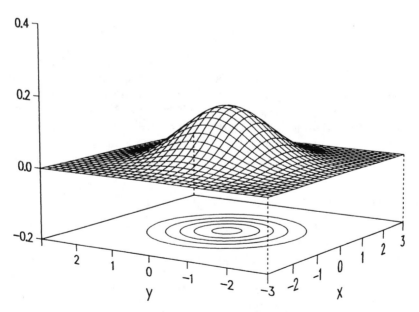

Figure 4.1. Bivariate normal $\rho = 0$.

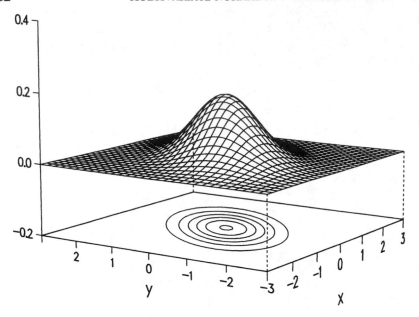

Figure 4.2. Bivariate normal $\rho = 0.5$.

Density Contours

For nonsingular Σ, the density function f has a constant value on the locus of points defined by $(x - \mu)'\Sigma^{-1}(x - \mu) = C$. This set of points corresponds to an ellipsoid, whose probability content can be deduced from a $\chi^2_{(p)}$ distribution. Some density plots (Figures 4.1–4.4) for the bivariate normal with standard normal marginals are presented, with the corresponding correlation ρ as indicated. Within each figure, the elliptical contours share the same major and minor axes, the diagonals $x = y$ and $x = -y$.

Variate Generation

The essence of a method for generating random vectors distributed $N_p(\mu, \Sigma)$ was alluded to earlier. Let X be $N_p(\mu, \Sigma)$ and Y be $N_p(0, I)$. If Σ is nonsingular and A is a $p \times p$ matrix such that $AA' = \Sigma$, then

$$X = AY + \mu, \tag{4.2}$$

where the equality is in distribution. Since Y can be generated by p successive calls to a univariate normal generator, (4.2) indicates the ap-

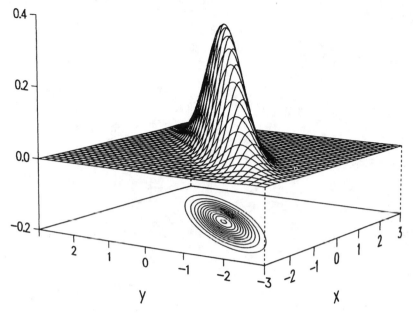

Figure 4.3. Bivariate normal $\rho = 0.9$.

propriate linear transformation to $\mathbf{Y}$ to achieve $N_p(\mu, \Sigma)$. The matrix A is not unique, however. For example, if Σ is a 2×2 correlation matrix,

$$\Sigma = \begin{bmatrix} 1 & \rho \\ \rho & 1 \end{bmatrix},$$

then both

$$A_1 = \begin{bmatrix} 1 & 0 \\ \rho & \sqrt{1 - \rho^2} \end{bmatrix}$$

and

$$A_2 = \begin{bmatrix} \sqrt{\left(\tfrac{1}{2}\right) + \left(\tfrac{1}{2}\right)\left(1 - \rho^2\right)^{1/2}} & \sqrt{\left(\tfrac{1}{2}\right) - \left(\tfrac{1}{2}\right)\left(1 - \rho^2\right)^{1/2}} \\ \sqrt{\left(\tfrac{1}{2}\right) - \left(\tfrac{1}{2}\right)\left(1 - \rho^2\right)^{1/2}} & \sqrt{\left(\tfrac{1}{2}\right) + \left(\tfrac{1}{2}\right)\left(1 - \rho^2\right)^{1/2}} \end{bmatrix}$$

yield $A_i A_i' = \Sigma$. Perhaps the best choice for A in (4.2) is provided by the Choleski factorization, which is the lower triangular matrix L for which

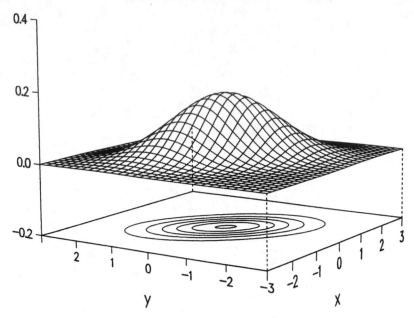

Figure 4.4. Bivariate normal $\rho = -0.5$.

$LL' = \Sigma$. This choice requires $p(p + 1)/2$ multiplications in $L\mathbf{X}$ versus p^2 in $A\mathbf{X}$. The Choleski factorization has the added advantage of being readily computable with simple recursion formulas. For up to three dimensions, the forms can be given compactly. Let

$$\Sigma = \begin{bmatrix} 1 & \rho_{12} & \rho_{13} \\ \rho_{12} & 1 & \rho_{23} \\ \rho_{13} & \rho_{23} & 1 \end{bmatrix}$$

be a correlation matrix. The appropriate matrix L is

$$L = \begin{bmatrix} 1 & 0 & 0 \\ \rho_{12} & \left(1 - \rho_{12}^2\right)^{1/2} & 0 \\ \rho_{13} & \dfrac{\rho_{23} - \rho_{12}\rho_{13}}{\left(1 - \rho_{12}^2\right)^{1/2}} & \dfrac{\left[\left(1 - \rho_{12}^2\right)\left(1 - \rho_{13}^2\right) - \left(\rho_{23} - \rho_{12}\rho_{13}\right)^2\right]^{1/2}}{\left(1 - \rho_{12}^2\right)^{1/2}} \end{bmatrix}.$$

For higher dimensions, subroutines in the widely distributed LINPACK (Dongarra et al., 1979) library can be used to obtain A.

The properties stated above will be frequently recounted to compare other multivariate distributions to the multivariate normal. In this sense, the multivariate normal distribution can serve as a baseline model in Monte Carlo work. Results obtained for the multivariate normal and a non-normal distribution can be interpreted with regard to characteristics of these distributions such as

1. Marginal distributions.
2. Conditional distributions—in particular, conditional means and co-variances.
3. Geometry of the density contours.
4. Appearance of the bivariate density function.
5. Inherent dependence as suggested by the correlation structure.
6. Location of the distribution, possibly indicated by a mean vector, if it exists.

For the multivariate distributions to be described, not all of the above characteristics will have slick analytical expressions. Some of these characteristics can be estimated via simulation and others can be gleaned from density contour plots for the bivariate distributions.

4.2. MIXTURES OF NORMAL VARIATES

A paradigm in the robustness literature is the contaminated normal distribution represented by

$$pN(\mu, \sigma^2) + (1 - p)N(\mu + \delta, k^2\sigma^2). \tag{4.3}$$

Here the addition refers to "mixing"—with probability p the process is realized from $N(\mu, \sigma^2)$; with probability $(1 - p)$ the process is realized from $N(\mu + \delta, k^2\sigma^2)$. For $\delta \neq 0$, the distribution is shifted in location from the baseline $N(\mu, \sigma^2)$. For $k \neq 1$, the distribution is scale contaminated. With appropriate choice of p, δ, and k, many commonly used statistical methods perform abominably with data generated from (4.3). Not surprisingly, the distribution in (4.3) is a popular choice for researchers who have devised robust statistical methods. Moreover, with five parameters present, many distributional shapes (including bimodal) are possible, some of which may even resemble actual data.

A useful paper in characterizing the density shape of (4.3) is by Eisenberger (1964). The distribution (4.3) cannot be bimodal if

$\delta^2 < 27k^2\sigma^2/[4(1 + k^2)]$. That is, every choice of p under this constraint produces a unimodal distribution. For $\delta^2 > 2k^2\sigma^2/(1 + k^2)$, a value of p exists for which (4.3) is bimodal.

In the multivariate case, we can mimic (4.3), yielding

$$pN_m(\mu_1, \Sigma_1) + (1 - p)N_m(\mu_2, \Sigma_2). \tag{4.4}$$

Generating variates from (4.4) is easy and can be accomplished as follows:

1. Generate U as uniform 0–1. If $U \leqslant p$, proceed to step 2. Otherwise, execute step 3.
2. Generate **X** as $N_m(\mu_1, \Sigma_1)$.
3. Generate **X** as $N_m(\mu_2, \Sigma_2)$.

The problem in using (4.4) is not in variate generation but in parameter selection. The number of parameters in (4.4) is $m^2 + 3m + 1$, which corresponds to $2m$ means, $2m$ variances, $m^2 - m$ correlations, and the mixing probability p.

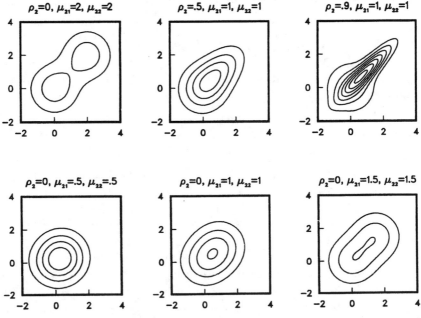

Figure 4.5. Bivariate normal mixtures, $p = 0.5$, $\rho_1 = 0$, $\mu_{11} = 0$, $\mu_{12} = 0$, $\sigma_{11} = 1$, $\sigma_{12} = 1$, $\sigma_{21} = 1$, $\sigma_{22} = 1$.

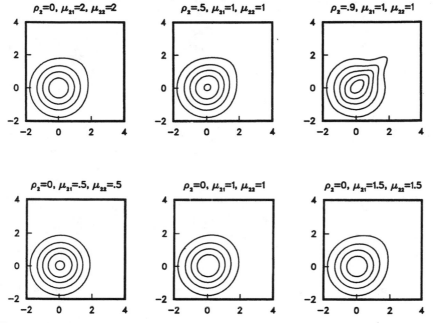

Figure 4.6. Bivariate normal mixtures, $p = 0.9$, $\rho_1 = 0$, $\mu_{11} = 0$, $\mu_{12} = 0$, $\sigma_{11} = 1$, $\sigma_{12} = 1$, $\sigma_{21} = 1$, $\sigma_{22} = 1$.

In the bivariate case, at least, a perusal of the shape of the density corresponding to (4.4) is attempted. The density function is

$$f(x, y) = pf_1(x, y) + (1 - p)f_2(x, y), \qquad (4.5)$$

where f_i is of the form (4.1), with means μ_{i1} and μ_{i2}, standard deviations σ_{i1} and σ_{i2}, and correlation ρ_i. Ten contour plots (Figures 4.5–4.14) are provided to indicate the range of appearance of (4.5).

It is useful to have expressions for the means and covariances of the distribution given in (4.4). We have if **X** is distributed as (4.4), then

$$E(\mathbf{X}) = p\mu_1 + (1 - p)\mu_2$$

$$\text{Cov}(\mathbf{X}) = p\Sigma_1 + (1 - p)\Sigma_2 + p(1 - p)\psi\psi',$$

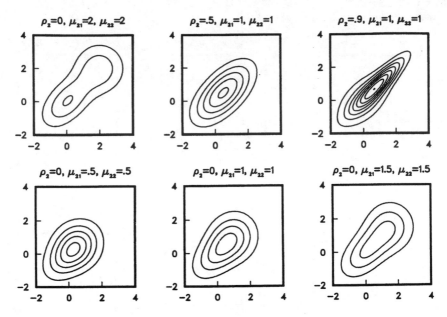

Figure 4.7. Bivariate normal mixtures, $p = 0.5$, $\rho_1 = 0.5$, $\mu_{11} = 0$, $\mu_{12} = 0$, $\sigma_{11} = 1$, $\sigma_{12} = 1$, $\sigma_{21} = 1$, $\sigma_{22} = 1$.

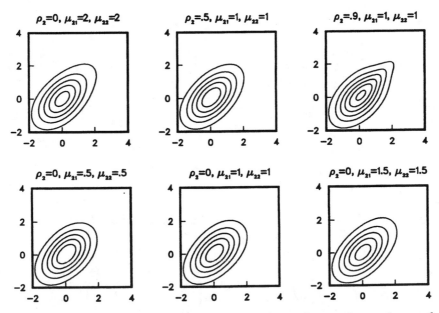

Figure 4.8. Bivariate normal mixtures, $p = 0.9$, $\rho_1 = 0.5$, $\mu_{11} = 0$, $\mu_{12} = 0$, $\sigma_{11} = 1$, $\sigma_{12} = 1$, $\sigma_{21} = 1$, $\sigma_{22} = 1$.

58

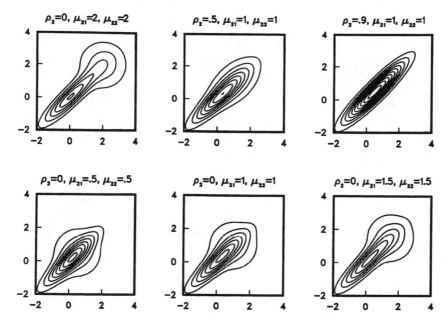

Figure 4.9. Bivariate normal mixtures, $p = 0.5$, $\rho_1 = 0.9$, $\mu_{11} = 0$, $\mu_{12} = 0$, $\sigma_{11} = 1$, $\sigma_{12} = 1$, $\sigma_{21} = 1$, $\sigma_{22} = 1$.

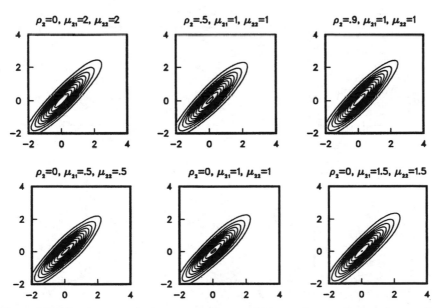

Figure 4.10. Bivariate normal mixtures $p = 0.9$, $\rho_1 = 0.9$, $\mu_{11} = 0$, $\mu_{12} = 0$, $\sigma_{11} = 1$, $\sigma_{12} = 1$, $\sigma_{21} = 1$, $\sigma_{22} = 1$.

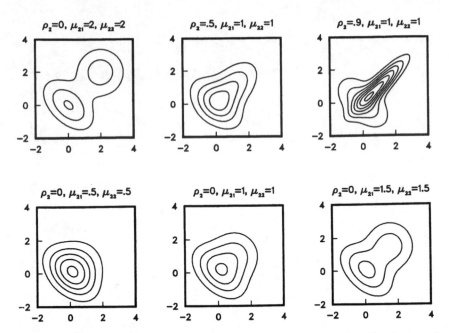

Figure 4.11. Bivariate normal mixtures, $p = 0.5$, $\rho_1 = -0.5$, $\mu_{11} = 0$, $\mu_{12} = 0$, $\sigma_{11} = 1$, $\sigma_{12} = 1$, $\sigma_{21} = 1$, $\sigma_{22} = 1$.

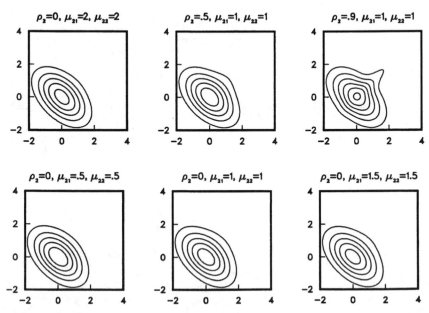

Figure 4.12. Bivariate normal mixtures, $p = 0.9$, $\rho_1 = -0.5$, $\mu_{11} = 0$, $\mu_{12} = 0$, $\sigma_{11} = 1$, $\sigma_{12} = 1$, $\sigma_{21} = 1$, $\sigma_{22} = 1$.

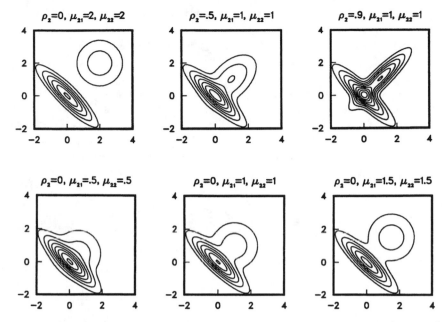

Figure 4.13. Bivariate normal mixtures, $p = 0.5$, $\rho_1 = -0.9$, $\mu_{11} = 0$, $\mu_{12} = 0$, $\sigma_{11} = 1$, $\sigma_{12} = 1$, $\sigma_{21} = 1$, $\sigma_{22} = 1$.

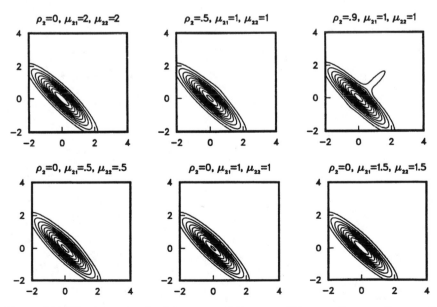

Figure 4.14. Bivariate normal mixtures, $p = 0.9$, $\rho_1 = -0.9$, $\mu_{11} = 0$, $\mu_{12} = 0$, $\sigma_{11} = 1$, $\sigma_{12} = 1$, $\sigma_{21} = 1$, $\sigma_{22} = 1$.

where

$$\psi = \mu_1 - \mu_2.$$

Obviously, in using (4.4) in a particular simulation application, only a limited number of cases can be considered. Perhaps a useful tactic is to expend the initial one-fourth of the resources in the bivariate setting, analyze the results, and proceed to higher dimensions with some reasonable restrictions on the parameters, gleaned from the bivariate investigation.

CHAPTER 5

Johnson's Translation System

The univariate Johnson translation system (Section 2.3) is readily extended to a multivariate system by applying the basic transformations to the individual components of a multivariate normal distribution (Johnson, 1949b). In particular, let $\mathbf{X}$ have an $N_p(\mu, \Sigma)$ distribution. To each component X_i of $\mathbf{X}$ apply one of the following transformations:

$$
\begin{array}{lll}
Y_i = X_i & \text{Normal} & \\
Y_i = \lambda_i \exp(X_i) + \xi_i & \text{Lognormal} & \\
Y_i = \lambda_i \sinh(X_i) + \xi_i & \text{Sinh}^{-1}\text{-normal} & \\
Y_i = \lambda_i [1 + \exp(X_i)]^{-1} + \xi_i & \text{Logit-normal} & (5.1)
\end{array}
$$

The resulting vector $\mathbf{Y} = (Y_1, Y_2, \ldots, Y_P)'$ has a distribution in the multivariate Johnson system. The simple forms of these functions lend themselves nicely to direct variate generation algorithms for the Johnson system. The only difficulty in employing the system in simulation work is to specify appropriate parameter combinations to meet the needs of particular applications. As an aid to distribution selection, a large collection of contour plots is provided in Sections 5.1–5.3. Although restricted to the bivariate case with the same basic distributional form (S_L, S_U, or S_B) for each component, these plots should be useful in the early stages of distribution selection. In terms of Johnson's shorthand notation, the distributions depicted are S_{LL}, S_{UU}, and S_{BB}. In general, the notation S_{IJ} refers to a pair (X_1, X_2) having component distributions S_I and S_J, respectively, where I and J can be N (normal), L (lognormal), U (sinh^{-1}-normal), or B (logit-normal).

Following a detailed discussion of the plots, some mathematical properties of the system are given. Of frequent interest in simulation applications
is the specification of the moment structure, for which two approaches are
indicated. Also covered are the conditional distributions in the bivariate
system. Finally, some previous discriminant analysis studies using the
Johnson system are reviewed. Some general guidelines in the conduct of
simulation studies emerge from this appraisal.

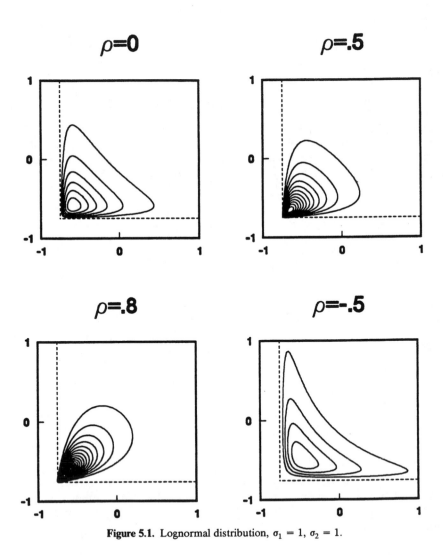

Figure 5.1. Lognormal distribution, $\sigma_1 = 1$, $\sigma_2 = 1$.

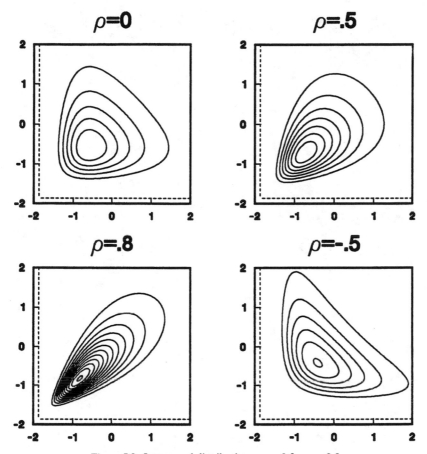

Figure 5.2. Lognormal distribution, $\sigma_1 = 0.5$, $\sigma_2 = 0.5$.

5.1.   PLOTS FOR THE S_{LL} DISTRIBUTION

The bivariate lognormal distribution presented in Figures 5.1–5.6 has the density function

$$g(x_1, x_2) = \frac{b_1 b_2}{2\pi\sigma_1\sigma_2(b_1 x_1 + a_1)(b_2 x_2 + a_2)(1 - \rho^2)^{1/2}}$$

$$\times \exp\left[-\frac{(y_1/\sigma_1)^2 - 2\rho y_1 y_2/(\sigma_1\sigma_2) + (y_2/\sigma_2)^2}{2(1 - \rho^2)}\right],$$

$$x_i > -a_i/b_i,\ i = 1, 2,$$

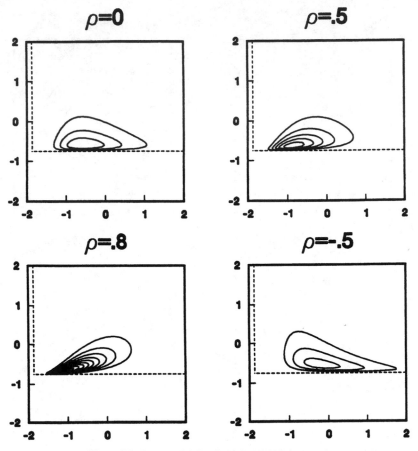

Figure 5.3. Lognormal distribution, $\sigma_1 = 0.5$, $\sigma_2 = 1$.

where

$$y_i = \ln(b_i x_i + a_i), \qquad\qquad i = 1, 2$$

$$a_i = \exp\left(\frac{\sigma_i^2}{2}\right), \qquad\qquad i = 1, 2$$

$$b_i = \left[\exp(2\sigma_i^2) - \exp(\sigma_i^2)\right]^{1/2} \qquad i = 1, 2.$$

This parameterization is advantageous for plotting purposes in that the

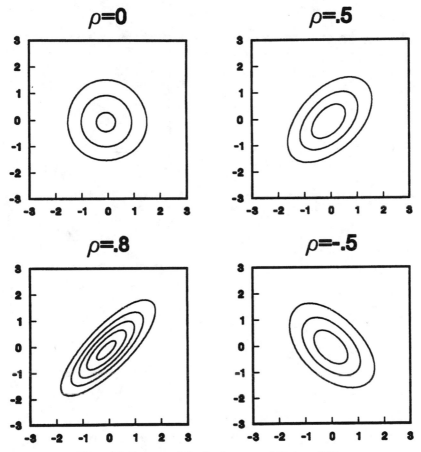

Figure 5.4. Lognormal distribution, $\sigma_1 = 0.05$, $\sigma_2 = 0.05$.

components have zero means and unit variances. This feature facilitates comparisons across the shape and dependence parameters of interest—σ_1, σ_2, and ρ. The parameters μ_1 and μ_2 are taken to be zero, since they are scale parameters in the lognormal distribution and λ_1 and λ_2 already serve that purpose. In terms of the transformation given by $Y_i = \lambda_i \exp(X_i) + \xi_i$,

$$\lambda_i = \frac{1}{b_i}$$

$$\xi_i = -\frac{a_i}{b_i}, \qquad i = 1, 2.$$

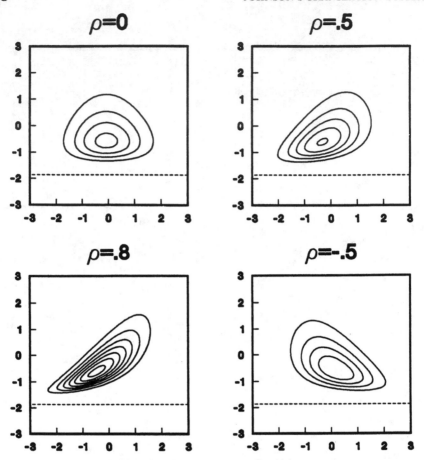

Figure 5.5. Lognormal distribution, $\sigma_1 = 0.05$, $\sigma_2 = 0.5$.

In each of the six sets of contour plots in Figures 5.1–5.6, four values of ρ are considered: 0.8, 0.5, 0.0, and -0.5. Three values of the shape parameter σ_i are considered ranging from 0.05, which gives an approximately normal component, to the intermediate value 0.5, and finally to 1.0, which corresponds to a very skewed distribution. Since the shape parameters σ_1 and σ_2 are allowed to differ for the two components, six distinct combinations of (σ_1, σ_2) result.

The dashed lines within a plot delineate the support of the distribution. For aesthetic reasons, the plots differ in terms of the range of the distribu-

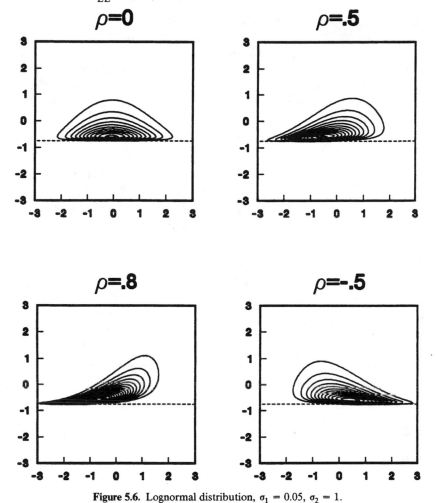

Figure 5.6. Lognormal distribution, $\sigma_1 = 0.05$, $\sigma_2 = 1$.

tion presented. The highly skewed cases have large values of the density at the mode so that different contour increments were used for the various plots.

The set of plots in Figure 5.4 for $(\sigma_1, \sigma_2) = (0.05, 0.05)$ look very similar to that expected for a bivariate normal—similar ellipses with common major and minor axes. As either or both σ_i's increase, substantial non-normality is apparent.

5.2. PLOTS FOR THE S_{UU} DISTRIBUTION

The bivariate $\sinh^{-1}$-normal distribution inherits five parameters u_1, u_2, σ_1, σ_2, and ρ from the bivariate normal distribution. The additional scale and location parameters λ_1, λ_2, ξ_1, and ξ_2 can be easily determined as functions of the μ_i's and σ_i's to yield components having zero means and unit variances. The density function whose contours appear in Figures

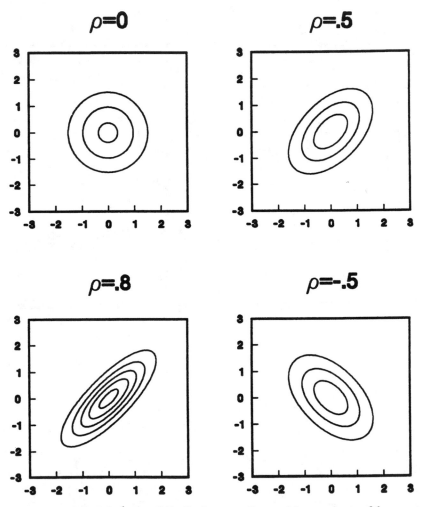

Figure 5.7. $\sinh^{-1}$-normal distribution, $\mu_1 = 0$, $\sigma_1 = 0.1$, $\mu_2 = 0$, $\sigma_2 = 0.1$.

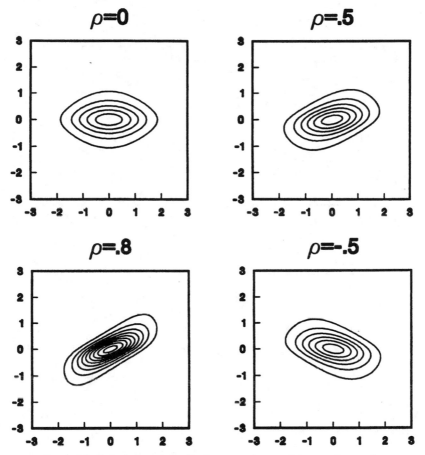

Figure 5.8. Sinh^{-1}-normal distribution, $\mu_1 = 0$, $\sigma_1 = 0.1$, $\mu_2 = 0$, $\sigma_2 = 1$.

5.7–5.16 is

$$f(x_1, x_2) = \frac{b_1 b_2 \left[w_1 + \left(1 + w_1^2\right)^{1/2} \right] \left[w_2 + \left(1 + w_2^2\right)^{1/2} \right]}{\left[1 + w_1^2 + w_1 \left(1 + w_1^2\right)^{1/2} \right] \left[1 + w_2^2 + w_2 \left(1 + w_2^2\right)^{1/2} \right]}$$

$$\times g\left[\sinh^{-1}(w_1), \sinh^{-1}(w_2) \right],$$

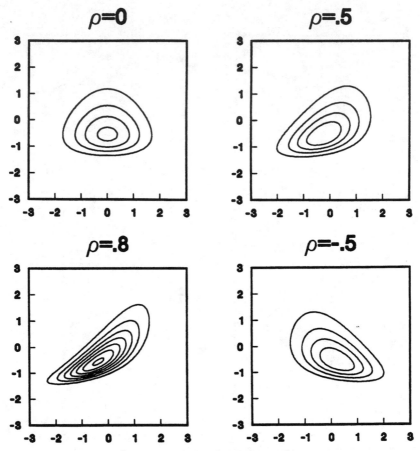

Figure 5.9. Sinh^{-1}-normal distribution, $\mu_1 = 0$, $\sigma_1 = 0.1$, $\mu_2 = 2$, $\sigma_2 = 0.5$.

where

$$w_i = b_i x_i + a_i$$

$$a_i = \exp\left(\frac{\sigma_i^2}{2}\right)\sinh(\mu_i)$$

$$b_i = \left\{(e^{\sigma_i^2} - 1)\left[e^{\sigma_i^2}\cosh(2\mu_i) + 1\right]\right\}^{1/2}, \qquad i = 1, 2$$

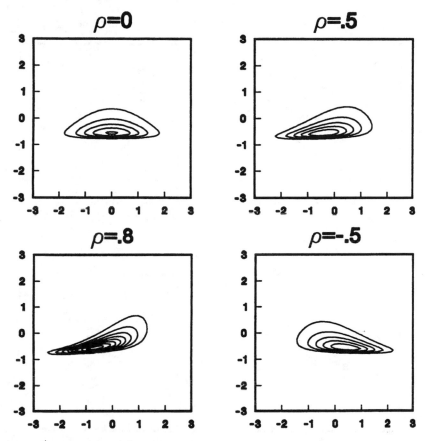

Figure 5.10. Sinh^{-1}-normal distribution, $\mu_1 = 0$, $\sigma_1 = 0.1$, $\mu_2 = 2$, $\sigma_2 = 1$.

$$g(y_1, y_2) = \frac{\exp\left\{-\left(y_1^2 - 2\rho y_1 y_2 + y_2^2\right)/\left[2\left(1 - \rho^2\right)\right]\right\}}{2\sigma_1\sigma_2\sqrt{1 - \rho^2}}$$

$$\sinh^{-1}(x) = \ln\left[x + \left(1 + x^2\right)^{1/2}\right].$$

From reviewing Figure 2.6 for the univariate S_U distribution, four reasonably distinct cases based on μ_i and σ_i were selected:

1. $\mu_i = 0$, $\sigma_i = 0.1$ Symmetric, nearly normal
2. $\mu_i = 0$, $\sigma_i = 1$ Symmetric, heavier tailed than normal
3. $\mu_i = 2$, $\sigma_i = 0.5$ Somewhat skewed
4. $\mu_i = 2$, $\sigma_i = 1$ Very skewed

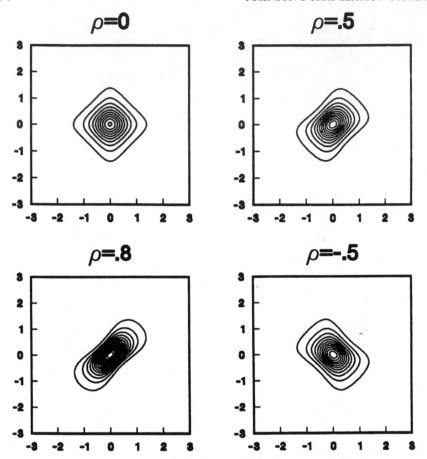

Figure 5.11. Sinh^{-1}-normal distribution, $\mu_1 = 0$, $\sigma_1 = 1$, $\mu_2 = 0$, $\sigma_2 = 1$.

These four component distributions give rise to ten combinations in the bivariate setting, where for each of the Figures 5.7–5.16, four values of the dependence parameter ρ are represented—0.8, 0.5, 0.0, and -0.5. From scrutiny of the plots, the following observations are made:

1. Figure 5.7 illustrates that the bivariate sinh^{-1}-normal can resemble the bivariate normal quite closely. A similar-looking plot would also be obtained (but is not presented here) for parameter value combinations:

   (a) $\mu_1 = \mu_2 = 0$, $\sigma_1 = \sigma_2 = 0.5$.
   (b) $\mu_1 = \mu_2 = 2$, $\sigma_1 = \sigma_2 = 0.1$.

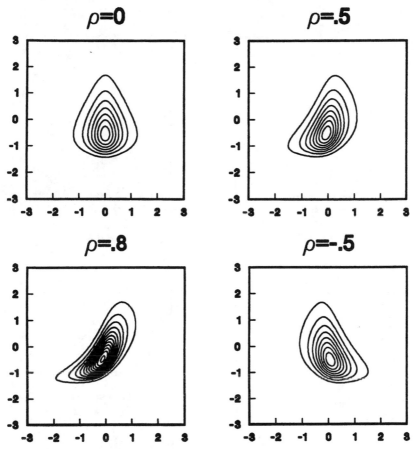

Figure 5.12. Sinh^{-1}-normal distribution, $\mu_1 = 0$, $\sigma_1 = 1$, $\mu_2 = 2$, $\sigma_2 = 0.5$.

2. A striking feature in Figure 5.8 is the tilt in the contours relative to the inner ellipse for the three nonzero ρ cases.

3. For $\mu_1 = \mu_2$ and $\sigma_1 = \sigma_2$ cases (Figures 5.7, 5.11, 5.14, 5.16), the densities are symmetric about the positive diagonal.

4. The three $\mu_1 = 0$, $\sigma_1 = 1$ plots (Figures 5.11–5.13) offer intriguing outer contour shapes ranging from baseball diamonds and near rectangles, through teardrops, to fluffy triangles and boomerangs. Such descriptors help to envisage the S_{UU} varieties.

5. The final three figures (5.14–5.16) ought to be reminiscent of their lognormal counterparts given earlier in Figures 5.4–5.6. For these parame-

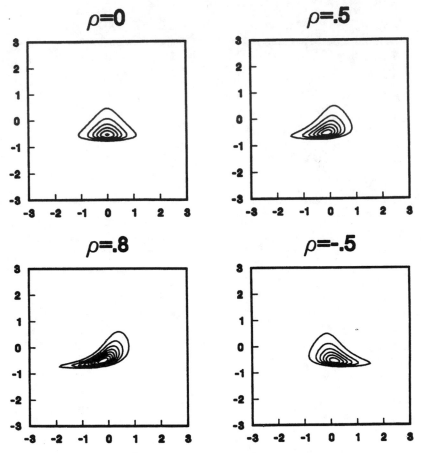

Figure 5.13. Sinh^{-1}-normal distribution, $\mu_1 = 0$, $\sigma_1 = 1$, $\mu_2 = 2$, $\sigma_2 = 1$.

ter choices, the sinh function is effectively an exponential function over the main probability content of an $N(2,1)$ distribution. Thus the transformed distribution should be approximately lognormal.

5.3.   CONTOUR PLOTS FOR THE S_{BB} DISTRIBUTION

The bivariate logit-normal distribution does not have a crisp standardized density form such that the components have zero means and unit variances. One advantage, however, for plotting purposes is that in the basic transfor-

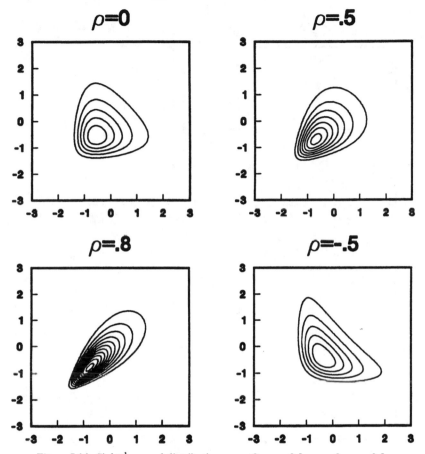

Figure 5.14. Sinh^{-1}-normal distribution, $\mu_1 = 2$, $\sigma_1 = 0.5$, $\mu_2 = 2$, $\sigma_2 = 0.5$.

mation $Y_i = \lambda_i[1 + \exp(X_i)]^{-1} + \xi_i$, the support of Y_i is the unit interval if $\lambda_i = 1$ and $\xi_i = 0$. Adopting this tack, the density for the S_{BB} distribution represented in Figures 5.17–5.37 is

$$g(x_1, x_2) = \frac{\exp\{-(y_1^2 - 2\rho y_1 y_2 + y_2^2)/[2(1 - \rho^2)]\}}{x_1 x_2(1 - x_1)(1 - x_2)2\pi\sigma_1\sigma_2(1 - \rho^2)^{1/2}},$$

where

$$y_i = \frac{\ln[x_i/(1 - x_i)] - \mu_i}{\sigma_i}, \qquad i = 1, 2.$$

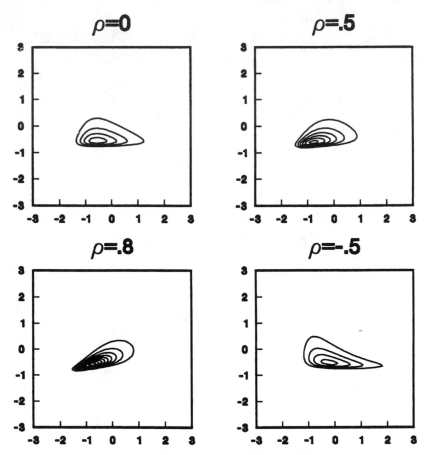

Figure 5.15. Sinh^{-1}-normal distribution, $\mu_1 = 2$, $\sigma_1 = 0.5$, $\mu_2 = 2$, $\sigma_2 = 1$.

The component distribution densities have a variety of shapes depending on μ_i and σ_i. For $\mu_i = 0$, the densities are symmetric about 0.5, with a mode or antimode at 0.5 depending on σ_i. For an extreme case, the density can be quite asymmetric with two local maxima (recall Figure 2.7). In view of the many possible univariate density shapes, a rather large assembly of bivariate contour plots is given. The following comments are intended to highlight some salient features from the S_{BB} plots:

1. Only 1 of the 21 cases roughly resembles the bivariate normal (Figure 5.17). This case has $\mu_1 = \mu_2 = 0$ and $\sigma_1 = \sigma_2 = 0.5$. Smaller values of the σ_i's would have improved the match between S_{BB} and S_{NN}. However, the

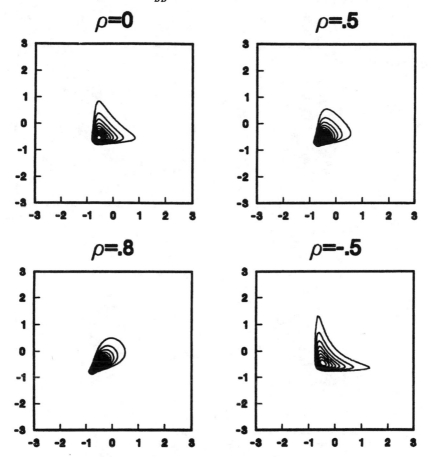

Figure 5.16. Sinh^{-1}-normal distribution, $\mu_1 = 2$, $\sigma_1 = 1$, $\mu_2 = 2$, $\sigma_2 = 1$.

corresponding plot would have used only a small fraction of the area of the unit square, since no standardizing was employed.

2. The component density shapes seem to have considerable influence on the shape of the joint density contours. This can be verified by examining first a case with a normal-like component, then modifying this component to be heavier tailed but symmetric, and then finally considering a bimodal component whose symmetric density's maxima are at zero and one. Such three-step progressions are given by the following:

(a) 5.20 → 5.21 → 5.22.

(b) 5.23 → 5.24 → 5.26.

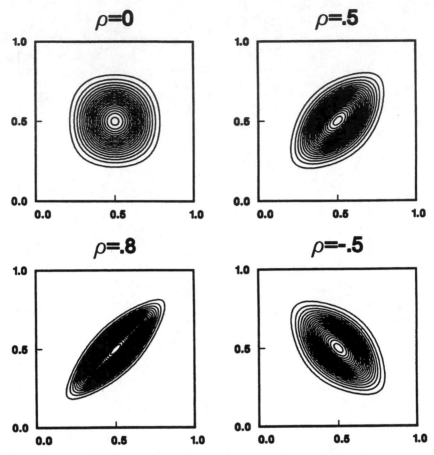

Figure 5.17. Logit-normal distribution, $(\mu_1, \sigma_1, \mu_2, \sigma_2) = (0, 0.5, 0, 0.5)$.

(c) $5.26 \rightarrow 5.27 \rightarrow 5.28$.

(d) $5.29 \rightarrow 5.30 \rightarrow 5.31$.

One way to understand what is going on here is to imagine a deformable plastic sheet that represents the density function at the first step in the progression. The sheet is held on each side and then pulled sideways, while preserving the unit volume beneath it. By the third step, the surface has been abutted against the support boundaries, giving rise to a mode at each side. A tricky aspect of this hypothetical exercise is envisioning the distribution of the second component as invariant to this stretching.

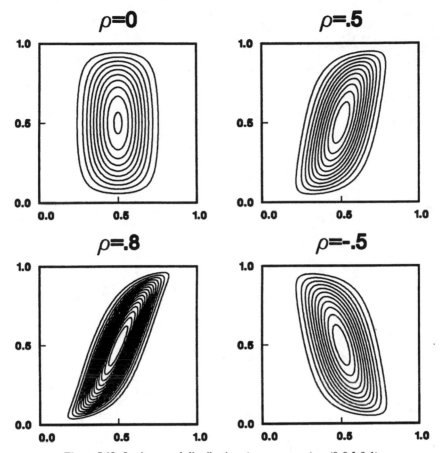

Figure 5.18. Logit-normal distribution, $(\mu_1, \sigma_1, \mu_2, \sigma_2) = (0, 0.5, 0, 1)$.

3. The process in item 2 can be continued to handle a change in the other component's distributions. Consider the following order in viewing the figures:

$$5.17 \to 5.18 \to \begin{matrix} 5.19 \\ \text{or} \\ 5.20 \end{matrix} \to 5.21 \to 5.22.$$

The density in Figure 5.22 has four modes, one at each corner.

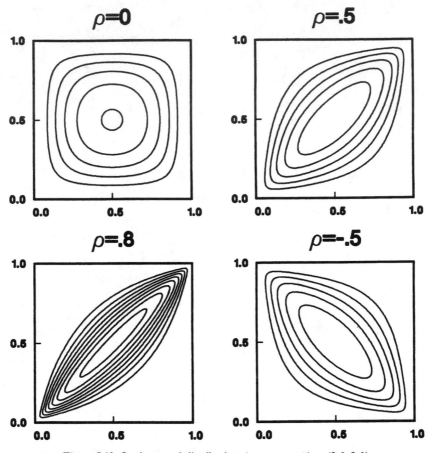

Figure 5.19. Logit-normal distribution, $(\mu_1, \sigma_1, \mu_2, \sigma_2) = (0, 1, 0, 1)$.

4. Strong dependence between components is reflected in the plots by mass concentrations (many contour lines jammed together) near the support boundaries—especially the corners. Good examples of this are Figure 5.31 and 5.37 with $\rho = 0.8$. The contour level increment was set at 0.5 for all S_{BB} plots in Figures 5.17–5.37.

5. Some of the more difficult-to-imagine density surfaces are depicted in Figures 5.38–5.42.

5.4. ANALYTICAL RESULTS

The preceding plots demonstrate that the Johnson system provides a diverse set of alternatives to the usual baseline distribution, the bivariate normal S_{NN}. These plots were restricted to distributions with common component distributions—S_{LL}, S_{UU}, and S_{BB}. For general S_{IJ} distributions, further variants in the plots could be obtained, although the number of cases would become exorbitant. In lieu of presenting more plots, some basic analytical results are provided for the Johnson system. Table 5.1 summarizes the key results on conditional distributions from which the corresponding conditional means and variances can be extracted. For example, suppose (X_1, X_2) is S_{UL}. The conditional distribution of X_1 given $X_2 = x_2$ is S_U with

TABLE 5.1. Summary of Formulas for the Johnson Translation System

Transformations

$$f_L(x) = \exp(x) \qquad f_L^{-1}(x) = \ln(x)$$
$$f_U(x) = \sinh(x) \qquad f_U^{-1}(x) = \sinh^{-1}(x) = \ln\left(x + \sqrt{x^2 + 1}\right)$$
$$f_B(x) = [1 + \exp(-x)]^{-1} \qquad f_B^{-1}(x) = \ln[y/(1 - y)]$$
$$f_N(x) = x \qquad f_N^{-1}(x) = x$$

Means and Variances, $X \sim N_1(\mu, \sigma^2)$

$$E[f_N(X)] = \mu$$
$$\text{Var}[f_N(X)] = \sigma^2$$
$$E[f_L(X)] = \exp(\mu + \sigma^2/2)$$
$$\text{Var}[f_L(X)] = [\exp(2\mu + \sigma^2)][\exp(\sigma^2) - 1]$$
$$E[f_U(X)] = \exp(\sigma^2/2)\sinh(\mu)$$
$$\text{Var}[f_U(X)] = [\exp(\sigma^2) - 1][1 + \cosh(2\mu)\exp(\sigma^2)]/2$$

Conditional Distributions

$$\begin{bmatrix} X_1 \\ X_2 \end{bmatrix} \sim N_2\left(\begin{bmatrix} \mu_1 \\ \mu_2 \end{bmatrix}, \begin{bmatrix} \sigma_1^2 & \rho\sigma_1\sigma_2 \\ \rho\sigma_1\sigma_2 & \sigma_2^2 \end{bmatrix}\right)$$
$$Y_1 = f_I(X_1), \qquad Y_2 = f_J(X_2)$$
Y_1 given $Y_2 = y_2$ is
$$f_I\left[N_1\left\{\mu_1 + (\rho\sigma_1/\sigma_2)[f_J^{-1}(y_2) - \mu_2], [1 - \rho^2]\sigma_1^2\right\}\right]$$

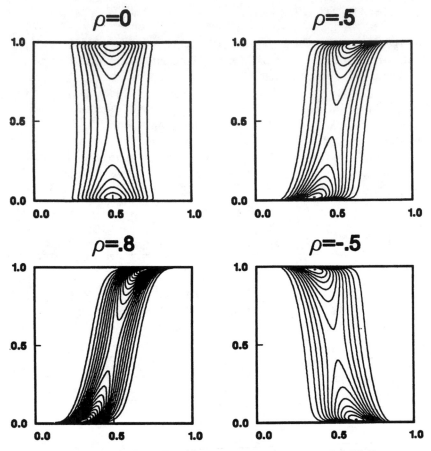

Figure 5.20. Logit-normal distribution, $(\mu_1, \sigma_1, \mu_2, \sigma_2) = (0, 0.5, 0, 2)$.

parameters

$$\mu = \mu_1 + \frac{\rho \sigma_1}{\sigma_2} \left[\ln(x_2) - \mu_2 \right]$$

$$\sigma^2 = \left(1 - \rho^2\right) \sigma_1^2.$$

Hence,

$$E(X_1 | X_2 = x_2) = \exp\left(\sigma_1^2 \frac{1 - \rho^2}{2} \right) \sinh\left\{ \mu_1 + \frac{\rho \sigma_1}{\sigma_2} \left[\ln(x_2) - \mu_2 \right] \right\}$$

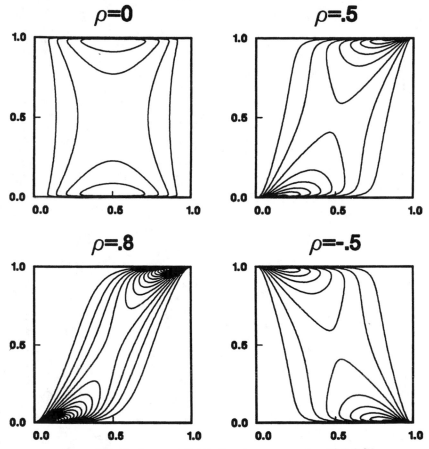

Figure 5.21. Logit-normal distribution, $(\mu_1, \sigma_1, \mu_2, \sigma_2) = ((0, 1, 0, 2)$.

and

$$\mathrm{Var}(X_1 | X_2 = x_2) = \tfrac{1}{2}\left\{\exp\left[(1 - \rho^2)\sigma_1^2\right] - 1\right\}$$

$$\times \left[1 + \cosh\left\{2\mu_1 + \frac{2\rho\sigma_1}{\sigma_2}\left[\ln(x_2) - \mu_2\right]\right\}\right.$$

$$\left. \times \exp\left[(1 - \rho^2)\sigma_1^2\right]\right].$$

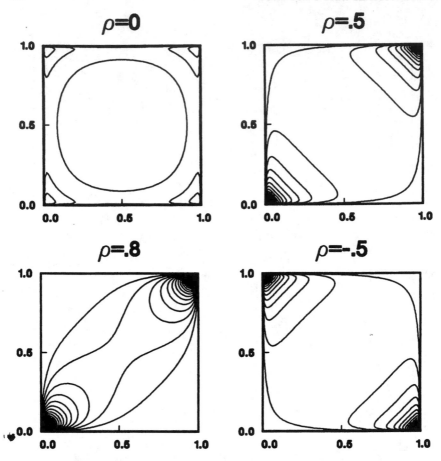

Figure 5.22. Logit-normal distribution, $(\mu_1, \sigma_1, \mu_2, \sigma_2) = (0, 2, 0, 2)$.

It could easily be expensive and difficult, if not overwhelming, to draw overall conclusions from a simulation study employing the Johnson system to its fullest extent—S_{IJ} for all of the I–J possibilities and an encompassing set of cases. The different supports of the components S_L, S_U, and S_B may make it difficult to justify mixing the component forms in many multivariate settings. At least initially it would seem prudent to restrict attention to distributions in the system having the same basic form of components.

In using the multivariate lognormal or multivariate $\sinh^{-1}$-normal distribution it can be helpful to specify mean vectors and covariance structures.

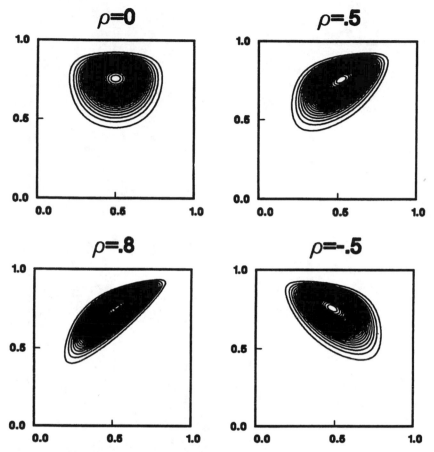

Figure 5.23. Logit-normal distribution, $(\mu_1, \sigma_1, \mu_2, \sigma_2) = (0, 0.5, 1, 0.5)$.

This facet of their use has played an important role in discriminant analysis simulation studies, as described later in this chapter. Specifying the first and second order moment structure in the multivariate $\sinh^{-1}$-normal distribution is quite easy, and will be considered first. Let $\mathbf{X}$ be $N_p(\mu, \Sigma)$ and define

$$\mathbf{Y} = (Y_1, Y_2, \ldots, Y_p)'$$

$$= [\sinh(X_1), \sinh(X_2), \ldots, \sinh(X_p)]'.$$

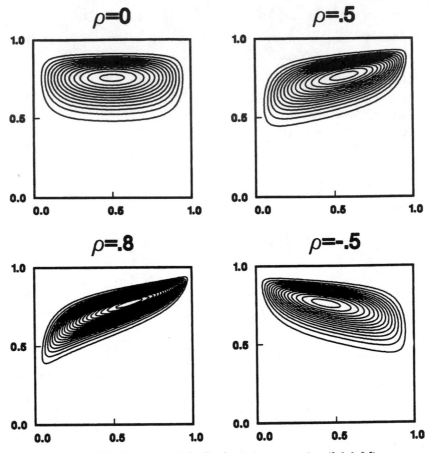

Figure 5.24. Logit-normal distribution, $(\mu_1, \sigma_1, \mu_2, \sigma_2) = (0, 1, 1, 0.5)$.

For notation let

$$E(Y_i) = \mu_i^*, \quad i = 1, 2, \ldots, p$$

$$\mathrm{Var}(Y_i) = \sigma_i^{*2}, \quad i = 1, 2, \ldots, p$$

$$\mathrm{Corr}(Y_i, Y_j) = \rho_{ij}^*, \quad i, j = 1, 2, \ldots, p,$$

and let μ_i, σ_i^2, and ρ_{ij} denote analogous quantities in the $N_p(\mu, \Sigma)$

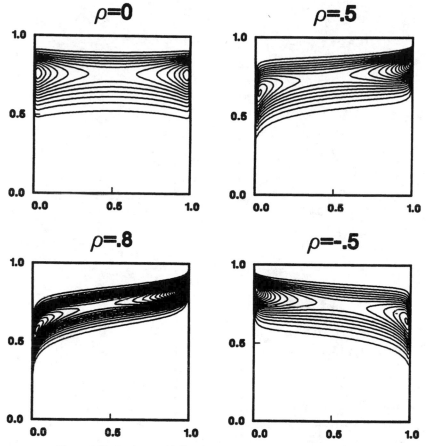

Figure 5.25. Logit-normal distribution, $(\mu_1, \sigma_1, \mu_2, \sigma_2) = (0, 2, 1, 0.5)$.

distribution. The following relationships hold:

$$\mu_i^* = \exp\left(\frac{\sigma_i^2}{2}\right)\sinh(\mu_i), \qquad\qquad i = 1, \ldots, p$$

$$\sigma_i^{*2} = \left[\exp(\sigma_i^2) - 1\right]\frac{1 + \cosh(2\mu_i)\exp(\sigma_i^2)}{2}, \qquad i = 1, \ldots, p$$

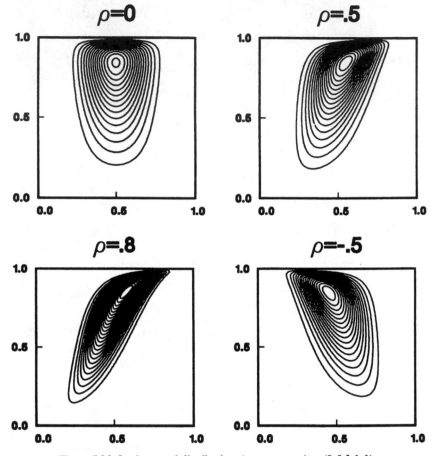

Figure 5.26. Logit-normal distribution, $(\mu_1, \sigma_1, \mu_2, \sigma_2) = (0, 0.5, 1, 1)$.

$$p^*_{ij} = (\sigma_i^* \sigma_j^*)^{-1} \exp\left(\frac{\sigma_i^2 + \sigma_j^2}{2}\right)\left[\exp(\rho_{ij}\sigma_i\sigma_j)\frac{\cosh(\mu_i + \mu_j)}{2}\right.$$

$$\left. - \exp(-\rho_{ij}\sigma_i\sigma_j)\frac{\cosh(\mu_i - \mu_j)}{2} - \sinh(\mu_i)\sinh(\mu_j)\right],$$

$$i, j = 1, \ldots, p, \quad (5.2)$$

where $\cosh(x) = [\exp(x) + \exp(-x)]/2$. These results can be obtained easily by recognizing that the kth moment of $Y_i = \sinh(X_i)$ is $[M_{X_i}(k) -$

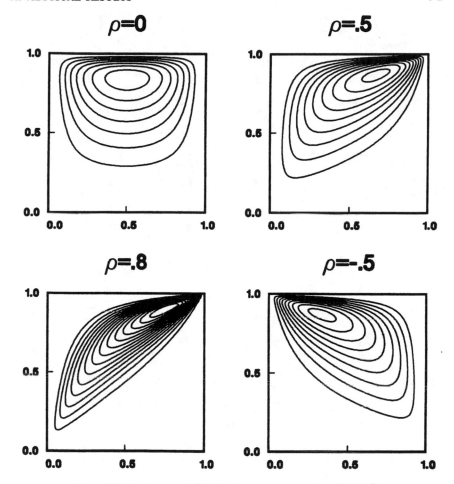

Figure 5.27. Logit-normal distribution, $(\mu_1, \sigma_1, \mu_2, \sigma_2) = (0, 1, 1, 1)$.

$M_{X_i}(-k)]/2$, where $M_{X_i}(t) = \exp(\mu t + \sigma^2 t^2 / 2)$ is the moment-generating function of a $N(\mu, \sigma^2)$ random variable.

Consider now the transformed Y_i given by

$$Z_i = \lambda_i \frac{Y_i - \mu_i^*}{\sigma_i^*} + \xi_i. \qquad (5.3)$$

The variate Z_i is $\sinh^{-1}$-normal with mean ξ_i and variance λ_i^2. Thus

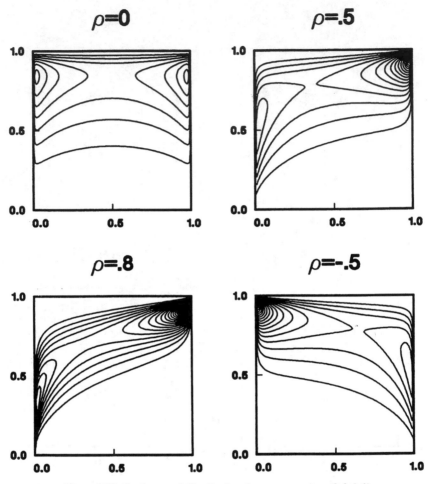

Figure 5.28. Logit-normal distribution, $(\mu_1, \sigma_1, \mu_2, \sigma_2) = (0, 2, 1, 1)$.

specifying component means and variances is straightforward. To specify correlations, the problem is to solve for ρ_{ij} in the third equation in (5.2). Since this equation is quadratic in the term $\exp(\rho_{ij}\sigma_i\sigma_j)$, we have the following:

$$\rho_{ij} = \frac{1}{\sigma_1\sigma_2}\ln\left[\frac{B + \left(B^2 + \cosh(\mu_1 + \mu_2)\cosh(\mu_1 - \mu_2)\right)}{\cosh(\mu_1 + \mu_2)}\right],$$

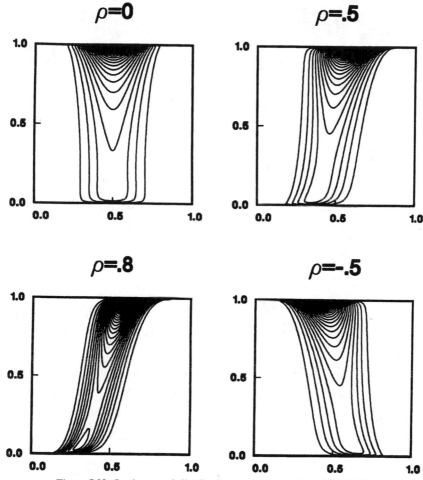

Figure 5.29. Logit-normal distribution, $(\mu_1, \sigma_1, \mu_2, \sigma_2) = (0, 0.5, 1, 2)$.

where $B = \sinh(\mu_1)\sinh(\mu_2) + \rho_{ij}^* \exp[-(\sigma_1^2 + \sigma_2^2)/2]$. Once the correct choice of individual correlations in the multivariate normal is made, some additional checks are required. If $|\rho_{ij}| > 1$ for some set (i, j), then the corresponding specified correlation ρ_{ij}^* is unobtainable. For three or higher dimensions, the required covariance matrix in the multivariate normal should also be tested for positive definiteness.

It should also be noted that in some cases it is possible to find values of the μ_i's and σ_i's alone to obtain specified means and variances in the

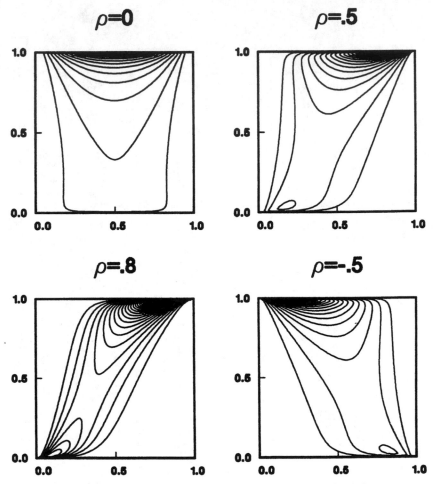

Figure 5.30. Logit-normal distribution, $(\mu_1, \sigma_1, \mu_2, \sigma_2) = (0, 1, 1, 2)$.

multivariate $\sinh^{-1}$-normal (Johnson, Ramberg, and Wang, 1982). This approach does not allow the introduction of the scale and location parameters in (5.3). Since this scheme effectively wastes the shape parameters on specifying scale and location values, the alternate approach embodied in (5.3) is much better.

An analogous approach to specifying means, variances, and correlations for the multivariate lognormal distribution can be developed. Let **X** be

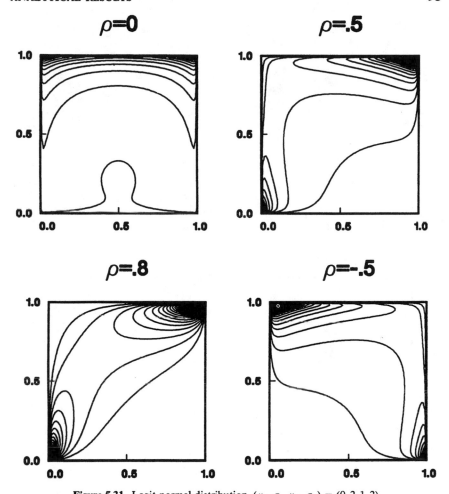

Figure 5.31. Logit-normal distribution, $(\mu_1, \sigma_1, \mu_2, \sigma_2) = (0, 2, 1, 2)$.

$N_p(\mathbf{0}, \dot{\Sigma})$ and set

$$\mathbf{Y} = \left(Y_1, Y_2, \ldots, Y_p\right)'$$
$$= \left[\exp(X_1), \exp(X_2), \ldots, \exp(X_p)\right]'.$$

The mean vector μ is set to $\mathbf{0}$ because control of scale in the lognormal

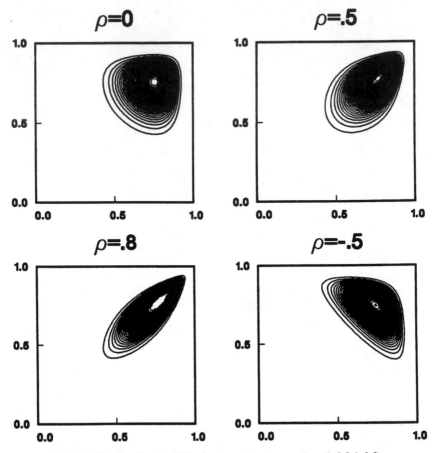

Figure 5.32. Logit-normal distribution, $(\mu_1, \sigma_1, \mu_2, \sigma_2) = (1, 0.5, 1, 0.5)$.

distribution can be handled directly in that realm. In particular, applying the transformation

$$Z_i = \lambda_i \frac{Y_i - \mu_i^*}{\sigma_i^*} + \xi_i,$$

where

$$\mu_i^* = \exp\!\left(\frac{\sigma_i^2}{2}\right)$$

$$\sigma_i^* = \left[\exp(2\sigma_i^2) - \exp(\sigma_i^2)\right]^{1/2},$$

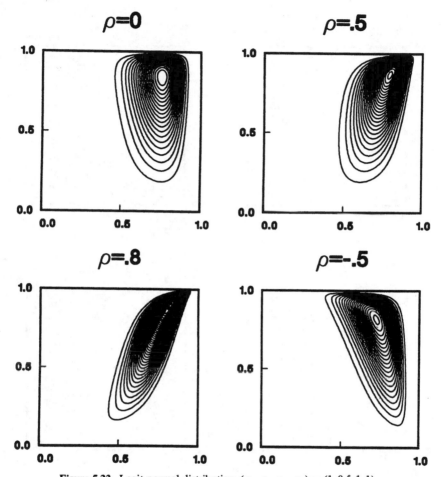

Figure 5.33. Logit-normal distribution, $(\mu_1, \sigma_1, \mu_2, \sigma_2) = (1, 0.5, 1, 1)$.

yields a lognormal variate Z_i having mean ξ_i and variance λ_i^2. The correlation ρ_{ij}^* between Y_i and Y_j is

$$\rho_{ij}^* = \frac{\exp(\rho_{ij}\sigma_i\sigma_j) - 1}{\left[\exp(\sigma_i^2) - 1\right]^{1/2}\left[\exp(\sigma_j^2) - 1\right]^{1/2}}.$$

Thus to obtain a specified correlation ρ_{ij}^* between Y_i and Y_j, the corre-

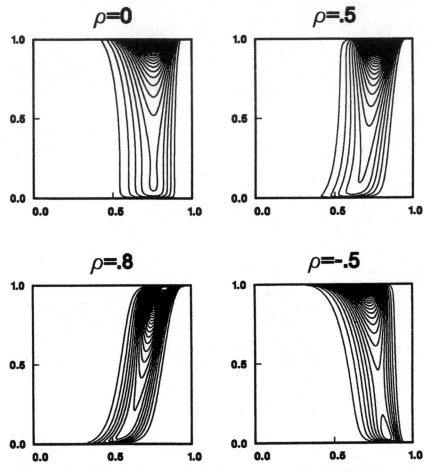

Figure 5.34. Logit-normal distribution, $(\mu_1, \sigma_1, \mu_2, \sigma_2) = (1, 0.5, 1, 2)$.

sponding correlation ρ_{ij} in the $N_p(0, \Sigma)$ distribution is found to be

$$\rho_{ij} = \frac{1}{\sigma_i \sigma_j} \ln \left\{ 1 + \rho_{ij}^* \left[\exp(\sigma_i^2) - 1 \right] \left[\exp(\sigma_j^2) - 1 \right] \right\}.$$

As in the $\sinh^{-1}$-normal moment specifications, it is possible that particular ρ_{ij}'s violate $|\rho_{ij}| \leqslant 1$ or that the ρ_{ij}'s in toto give rise to a matrix Σ that is

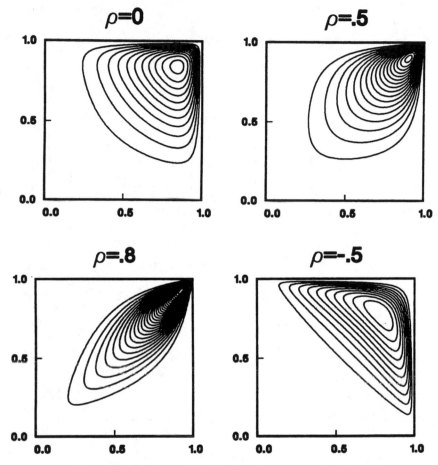

Figure 5.35. Logit-normal distribution, $(\mu_1, \sigma_1, \mu_2, \sigma_2) = (1, 1, 1, 1)$.

not positive definite. Either situation demands a reevaluation of the specified covariance matrix in the $\sinh^{-1}$-normal distribution.

5.5. DISCRIMINANT ANALYSIS APPLICATIONS

Several Monte Carlo studies in discriminant analysis have used the members of the Johnson translation system to model the distributions governing populations (see Section 1.2 for a discussion of related discriminant analysis

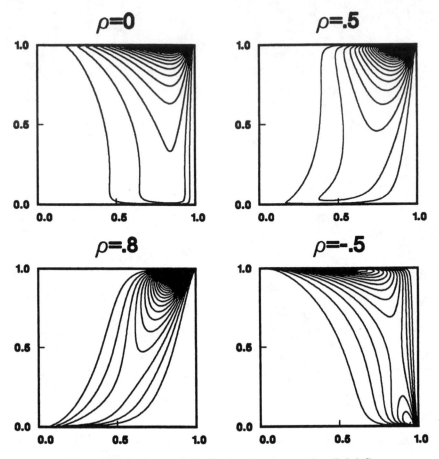

Figure 5.36. Logit-normal distribution, $(\mu_1, \sigma_1, \mu_2, \sigma_2) = (1, 1, 1, 2)$.

issues). One of the first papers was by Lachenbruch, Sneeringer, and Revo (1973), who studied the estimation of error rates in linear and quadratic discrimination under departures from the multivariate normal distribution. Their study provided the distributional framework for subsequent Monte Carlo work by Koffler and Penfield (1978). The specific distributions used from the Johnson system were four- and ten-dimensional multivariate distributions with components having the same basic form—lognormal, $\sinh^{-1}$-normal, or logit-normal. In the two-population discrimination problem, they started with two multivariate normal distributions with independent components and unit variances but with mean vectors for the

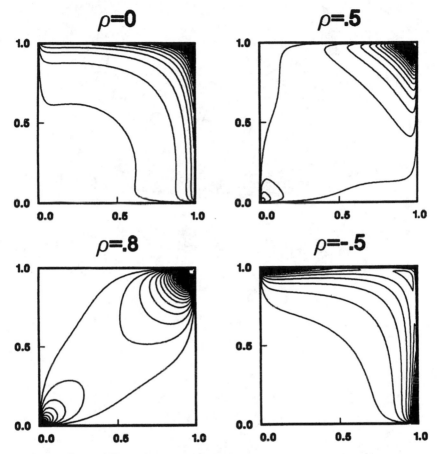

Figure 5.37. Logit-normal distribution, $(\mu_1, \sigma_1, \mu_2, \sigma_2) = (1, 2, 1, 2)$.

populations being different. In particular, the mean vector for population one (Π_1) was $(\delta, 0, \ldots, 0)$; for population two (Π_2) it was $(0, 0, \ldots, 0)$. The values of δ used were 1, 2, and 3. The transformed log-normal and $\sinh^{-1}$-normal distributions have variances as follows:

δ	$\mathrm{Var}(e^X)$	$\mathrm{Var}[\sinh(X)]$
0	5.67	3
1	34	9
2	255	74
3	1884	471

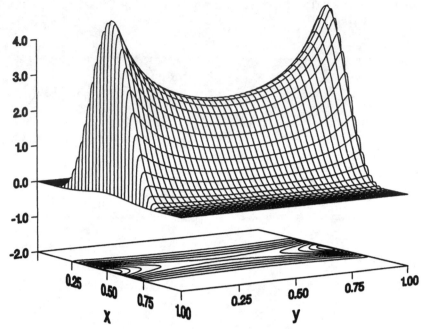

Figure 5.38. Density surface for the logit-normal distribution of Figure 5.20 with $\rho = 0$.

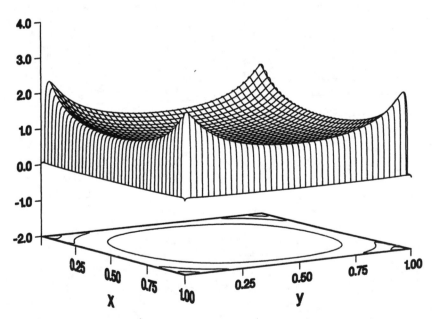

Figure 5.39. Density surface for the logit-normal distribution of Figure 5.22 with $\rho = 0$.

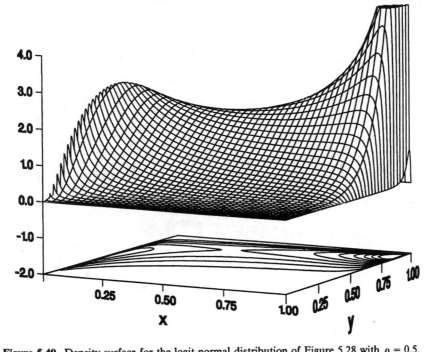

Figure 5.40. Density surface for the logit-normal distribution of Figure 5.28 with $\rho = 0.5$.

For δ equal to 3, the covariance matrices for Π_1 and Π_2 for the four-dimensional lognormal cases are

$$\Sigma_1 = \begin{bmatrix} 1884 & 0 & 0 & 0 \\ 0 & 5.67 & 0 & 0 \\ 0 & 0 & 5.67 & 0 \\ 0 & 0 & 0 & 5.67 \end{bmatrix}$$

$$\Sigma_2 = \begin{bmatrix} 5.67 & 0 & 0 & 0 \\ 0 & 5.67 & 0 & 0 \\ 0 & 0 & 5.67 & 0 \\ 0 & 0 & 0 & 5.67 \end{bmatrix}.$$

The main point of these calculations is to reveal that the two populations have rather different covariance structures. Since linear and quadratic discriminant procedures are also known to be degraded by unequal covariance structures in the populations, the design of Lachenbruch and colleagues

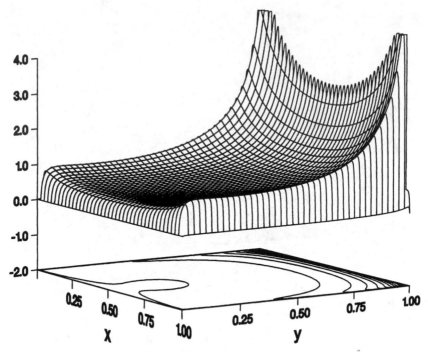

Figure 5.41. Density surface for the logit-normal distribution of Figure 5.31 with $\rho = 0$.

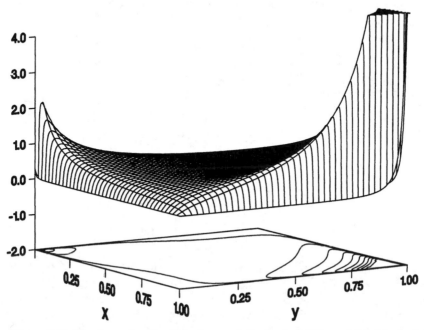

Figure 5.42. Density surface for the logit-normal distribution of Figure 5.37 with $\rho = 0.5$.

confounds the effects of non-normality and covariance structure. Although the authors state clearly that they did not control the covariances in the transformed populations, most papers citing their work seem to have overlooked the disclaimer (e.g., Krzanowski, 1977). The results given earlier in this section can be used to design a Monte Carlo study that avoids the confounding problem. Johnson, Ramberg, and Wang (1982) redid parts of the Lachenbruch study and concluded that non-normality as found in the Johnson translation system does degrade linear discrimination performance but not to the same extreme extent as reported by the previous authors. Despite this criticism, Lachenbruch, Sneeringer, and Revo should be credited with popularizing the use of the Johnson translation system in multivariate Monte Carlo discriminant analysis studies. Their work was certainly beneficial to subsequent Monte Carlo studies by Conover and Iman (1980) and Beckman and Johnson (1981).

CHAPTER 6

Elliptically Contoured Distributions

Elliptically contoured distributions provide a useful class of distributions for assessing the robustness of statistical procedures to certain types of multivariate non-normality. A statistical method is generally considered to be "robust" if it performs satisfactorily outside the set of assumptions on which it is based. The multivariate normal distribution is an elliptically contoured distribution sharing many but not all of its properties with the entire class. The multivariate Pearson Types II and VII distributions are two other special cases that are attractive in Monte Carlo work since they are easy to generate and cover a wide range within the class. Section 6.1 outlines basic properties of elliptically contoured distributions. Some particular distributions are examined in Section 6.2.

6.1. GENERAL RESULTS FOR ELLIPTICALLY CONTOURED DISTRIBUTIONS

Elliptically contoured distributions have received considerable attention recently in the statistical literature in both theoretical and applied contexts. On the theory side, elliptically contoured distributions are of interest since many results for the multivariate normal distribution carry over in this broader class. More pragmatically, experimental findings can often be conveniently summarized with models based on elliptically contoured distributions (e.g., the scattering distribution of particles from linear accelerators). Chmielewski (1981) provides an annotated bibliography describing the virtues of elliptically contoured distributions. Results for these distributions are also being incorporated prominently in multivariate analysis textbooks (Mardia, Kent, and Bibby, 1979 and Muirhead, 1982).

Elliptically contoured distributions can be defined in terms of the subclass of spherically symmetric distributions. A $p \times 1$ random vector X is spherically symmetric if the distribution of X is the same as PX for all $p \times p$ orthogonal matrices P. Geometrically, this definition asserts that spherically symmetric distributions are invariant under rotations. The broader class of elliptically contoured distributions can then be obtained through affine transformations of spherically symmetric distributions. Although providing an intuitive feel for these distributions, this definition lacks an explicit parametric framework for stating key results. The following definition as given by Cambanis, Huang, and Simons (1981) overcomes this difficulty.

Definition

Let X be a $p \times 1$ random vector, μ be a $p \times 1$ vector in R^p, and Σ be a $p \times p$ non-negative definite matrix. X has an elliptically contoured distribution if the characteristic function $\phi_{X-\mu}(t)$ of $X - \mu$ is a function of the quadratic form $t'\Sigma t$ as $\phi_{X-\mu}(t) = \phi(t'\Sigma t)$. This form can be written as

$$\phi_{X-\mu}(t) = \exp(it'\mu)\psi(t'\Sigma t) \tag{6.1}$$

for some function ψ.

In virtually all Monte Carlo applications it is reasonable to further restrict attention to elliptically symmetric distributions having density functions and nonsingular Σ. This restricted class has density functions of the form

$$f(x) = k_p|\Sigma|^{-1/2}g\left[(x - \mu)'\Sigma^{-1}(x - \mu)\right], \tag{6.2}$$

where g is a one-dimensional real-valued function independent of p and k_p is a scalar proportionality constant. For the multivariate normal $N_p(\mu, \Sigma)$ distribution, $g(t) = \exp(-t/2)$ and $k_p = (2\pi)^{-p/2}$. In one dimension, the class of elliptically contoured distributions consists of all symmetric distributions. Some basic properties and results for elliptically contoured distributions are now stated.

Notation

Elliptically contoured distributions as in (6.2) are denoted $EC_p(\mu, \Sigma; g)$. Spherically symmetric distributions are consequently denoted $EC_p(0, I; g)$

where $\mathbf{0}$ is the $p \times 1$ zero vector and I is the $p \times p$ identity matrix. If it exists, $E(\mathbf{X}) = \mu$.

Independence and Correlation

If $\mathbf{X}$ is $EC_p(\mu, \Sigma; g)$ and $\mathbf{X}$ has independent components, then $\mathbf{X}$ must be $N_p(\mu, \Sigma)$ with Σ a diagonal matrix. The converse is false: if Σ is a diagonal matrix, $\mathbf{X}$ is not necessarily multivariate normal. If the second order moments of $\mathbf{X}$ distributed as $EC_p(\mu, \Sigma; g)$ exist, then the correlation between two components, X_i and X_j, is $\alpha_{ij}/\sqrt{\alpha_{ii}\alpha_{jj}}$, where α_{ij} is the (i, j)th element of Σ. The matrix Σ in (6.1) is proportional but not necessarily equal to the covariance matrix of $\mathbf{X}$. In particular, $\mathrm{Cov}(\mathbf{X}) = a\Sigma$ where $a = -2\psi'(0)$, with the function ψ from (6.1).

Linear Forms

If $\mathbf{X} \sim EC_p(\mu, \Sigma; g)$ and B is an $r \times p$ matrix of rank r ($r \leqslant p$), then

$$BX \sim EC_r(B\mu, B\Sigma B'; g).$$

This result gives the mechanism to relate elliptical and spherical distributions directly. Let $\mathbf{Y} \sim EC_p(0, I; g)$ and suppose $\Sigma = AA'$ for some $p \times p$ matrix A. We have

$$\mathbf{X} = A\mathbf{Y} + \mu \sim EC_p(\mu, \Sigma; g). \tag{6.3}$$

Thus for generation purposes, it is sufficient to generate the spherical distributions, and then apply a form of (6.3).

Quadratic Forms

Let $\mathbf{X} \sim EC_p(\mu, \Sigma; g)$ with a density function as in (6.2). The density of the random variable $Z = (\mathbf{X} - \mu)'\Sigma^{-1}(\mathbf{X} - \mu)$ is

$$h(z) = \frac{\pi^{p/2}}{\Gamma(p/2)} k_p z^{p/2-1} g(z). \tag{6.4}$$

As noted in Section 4.1, for the case of the multivariate normal distribution h is the $\chi^2_{(p)}$ density function.

Directional Distributions

Suppose $\mathbf{X} \sim EC_p(0, I; g)$. The distribution of $\lambda\mathbf{X}$ where $\lambda = (X_1^2 + X_2^2 + \cdots + X_p^2)^{-1/2}$ is uniformly distributed on the boundary of the p-dimensional unit hypersphere. This distribution does not depend on g, the particular functional form of the spherically symmetric distribution.

Marginal Distributions

All marginal distributions of elliptically symmetric distributions are elliptically symmetric and have the same functional form. Explicitly, let $X \sim EC_p(\mu, \Sigma; g)$ and partition the arrays as

$$X = \begin{bmatrix} X_1 \\ X_2 \end{bmatrix} \quad \text{and} \quad \mu = \begin{bmatrix} \mu_1 \\ \mu_2 \end{bmatrix} \quad \text{with dimensions} \quad \begin{bmatrix} k \times 1 \\ (p - k) \times 1 \end{bmatrix}$$

$$\Sigma = \begin{bmatrix} \Sigma_{11} & \Sigma_{12} \\ \Sigma_{21} & \Sigma_{22} \end{bmatrix} \text{with dimensions} \begin{bmatrix} k \times k & k \times (p - k) \\ (p - k) \times k & (p - k) \times (p - k) \end{bmatrix}.$$

The distribution of X_1 is $EC_k(\mu_1, \Sigma_{11}; g)$ and $X_2 \sim EC_{p-k}(\mu_2, \Sigma_{22}; g)$. The density function of X_1, for example, is proportional to

$$g\left[(x_1 - \mu_1)'\Sigma_{11}^{-1}(x_1 - \mu_1)\right].$$

Conditional Distributions

Let X be partitioned as above. The conditional distribution of X_1 given $X_2 = x_2$ is elliptically contoured with location and scale parameters that do not depend upon the function g. If the conditional moments of X_1 given $X_2 = x_2$ exist, they are given by

$$E(X_1|X_2 = x_2) = \mu_1 + \Sigma_{12}\Sigma_{22}^{-1}(x_2 - \mu_2)$$

and

$$\text{Cov}(X_1|X_2 = x_2) = w(x_2)\left(\Sigma_{11} - \Sigma_{12}\Sigma_{22}^{-1}\Sigma_{21}\right),$$

where w is a function of x_2 through the quadratic form $(x_2 - \mu_2)'\Sigma_{22}^{-1}$ $(x_2 - \mu_2)$. For some cases of practical interest the function w has a simple form. Cambanis, Huang, and Simons (1981) give the general form of w and proofs of these results. These authors also provide a representation of elliptically contoured distributions useful for variate generation, which is considered next.

Variate Generation

If X is $EC_p(\mu, \Sigma; g)$, then it can be generated as

$$X = RBU^{(p)} + \mu,   \tag{6.5}$$

where R is a positive random variable independent of $U^{(p)}$ having the distribution of $[(X - \mu)'\Sigma^{-1}(X - \mu)]^{1/2}$, B is a $p \times p$ matrix such that $BB' = \Sigma$, and $U^{(p)}$ is a random vector that is uniformly distributed on the unit hypersphere. Methods for generating the uniform distribution $U^{(p)}$ are given in Chapter 7. Generation of R depends on the particular elliptically contoured distribution under consideration. Of course, the formula in (6.4) giving the density of R^2 facilitates this task. For several special cases described in Section 6.2, the distribution of R is easy to generate.

The representation given in (6.5) also offers a convenient way to determine if the moments of $X \sim EC_p(\mu, \Sigma; g)$ exist. The components of X have finite kth moment if and only if $E(R^k)$ is finite.

For additional discussion of elliptically contoured distributions, the following papers may be of interest: Lord (1954), McGraw and Wagner (1968), Kelker (1970), Das Gupta et al. (1972), Chu (1973), Higgins (1975), and Devlin, Gnanadesikan and Kettenring (1976). The annotated bibliography by Chmielewski (1981) is a useful guide to the literature. For multivariate simulation purposes, the elliptically contoured Pearson Type II and Type VII distributions covered next may be adequate to consider.

6.2. SPECIAL CASES OF ELLIPTICALLY CONTOURED DISTRIBUTIONS

The best known case of an elliptically contoured distribution is the multivariate normal distribution, whose properties were reported in Section 4.1. Some other elliptically contoured distributions are described in this section, with an emphasis on suitable candidates for Monte Carlo applications. Two particularly appealing families are the Pearson Types II and VII multivariate distributions. These distribution are easy to generate, have simple density functions, and possess properties that provide a useful contrast to the multivariate normal distribution. Following a detailed discussion of these two special cases, some miscellaneous elliptically contoured distributions are surveyed.

Pearson Type II Distribution

Kotz (1975) defines the p-dimensional Pearson Type II distribution with the density function

$$f(\mathbf{x}) = \frac{\Gamma(p/2 + m + 1)}{\Gamma(m + 1)\pi^{p/2}}|\Sigma|^{-1/2}\big[1 - (\mathbf{x} - \mu)'\Sigma^{-1}(\mathbf{x} - \mu)\big]^m, \quad (6.6)$$

having support $(\mathbf{x} - \mu)'\Sigma^{-1}(\mathbf{x} - \mu) \leqslant 1$ and shape parameter $m > -1$. Further, the parameters μ and Σ are interpreted as:

$$E(\mathbf{X}) = \mu$$

$$\text{Cov}(\mathbf{X}) = \frac{1}{2m + p + 2}\Sigma.$$

To provide a feel for the impact of the parameter m on the distribution given by (6.6), some density plots are given in Figures 6.1–6.7. These

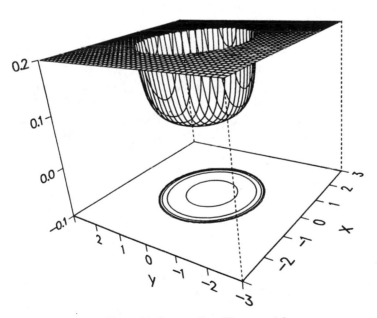

Figure 6.1. Pearson Type II, $m = -0.5$.

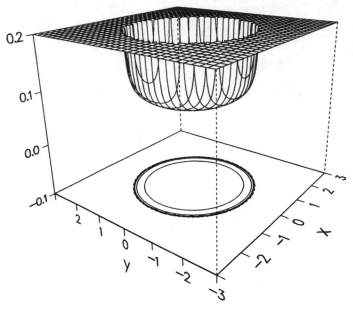

Figure 6.2. Pearson Type II, $m = -0.25$.

bivariate distributions have Σ selected as a diagonal matrix with identical nonzero entries of $2m + 4$. Thus comparisons across the seven values of $m = -0.5, -0.25, 0, 0.5, 1, 2,$ and 4 relate to distributions with the same marginal variances. The dashed line in the contour plane provides the boundary of the support of the distribution. The approach of the distribution to bivariate normality with increasing m is particularly striking in this progression of plots.

The form given in (6.6) can be used directly to obtain distributional characteristics. For example, let $\mathbf{X}$ have density (6.6) with $\mu = \mathbf{0}$ and $\Sigma = I$. The marginal distribution of $(X_1, X_2, \ldots, X_{p-1})'$ is derived by direct integration:

$$f(x_1, x_2, \ldots, x_{p-1})$$

$$= 2 \int_0^{(1 - x_1^2 - \cdots - x_{p-1}^2)^{1/2}} \frac{\Gamma(p/2 + m + 1)}{\Gamma(m + 1)\pi^{p/2}} \left(1 - x_1^2 - \cdots - x_p^2\right)^m dx_p$$

$$= \frac{2\Gamma(p/2 + m + 1)}{\Gamma(m + 1)\pi^{p/2}} \int_0^1 \frac{(1 - y)}{2\sqrt{y}} \left(1 - x_1^2 - \cdots - x_{p-1}^2\right)^{m+1/2} dy,$$

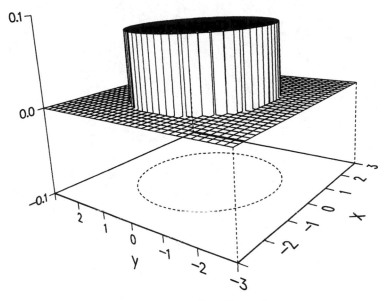

Figure 6.3. Pearson Type II, $m = 0$.

where the change of variable $y = x_p^2/(1 - x_1^2 - \cdots - x_{p-1}^2)$ has been made,

$$= \frac{\Gamma\left[(p-1)/2 + \left(m + \frac{1}{2}\right) + 1\right]}{\Gamma\left[\left(m + \frac{1}{2}\right) + 1\right]\pi^{(p-1)/2}}\left(1 - x_1^2 - \cdots - x_{p-1}^2\right)^{m+1/2}.$$

This form can be recognized by comparison with (6.6) to be a Pearson Type II density with shape parameter $m + \frac{1}{2}$. More generally, if $\mathbf{X}$ has density (6.6), the marginal distribution of any k components of $\mathbf{X}$ is Pearson Type II with shape parameter $m + (p - k)/2$ and with mean vector and scaling matrix extracted from μ and Σ. Of particular interest may be the marginal distribution of a single component X_i of $\mathbf{X}$. Assuming $\mu = 0$, $\Sigma = I$, the density of X_i is

$$f(x_i) = \frac{\Gamma(m + p/2 + 1)}{\Gamma\left(m + p + \frac{1}{2}\right)\Gamma\left(\frac{1}{2}\right)}\left(1 - x_i^2\right)^{m+(p-1)/2}, \qquad -1 \leqslant x_i \leqslant 1, \quad (6.7)$$

for $i = 1, 2, \ldots, p$. The marginal distribution obviously depends on the

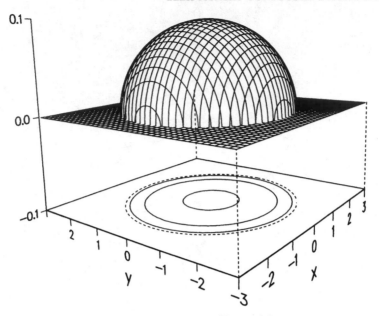

Figure 6.4. Pearson Type II, $m = 0.5$.

dimension p of **X**. The limiting marginal distributions as $p \to \infty$ and/or $m \to \infty$ are normal. Even in the "uniform" case ($m = 0$), the marginal distributions approach normality as the dimension increases. The univariate distribution in (6.7) is symmetric about zero, the variance is $(2m + p + 2)^{-1}$, and all of its moments exist. The kurtosis of X_i is $3(m + p/2 + 1)/(m + p/2 + 2)$, which converges monotonically to 3, the kurtosis of the normal distribution, as $m \to \infty$ or $p \to \infty$.

Conditional distributions are easy to derive for the case $\mu = \mathbf{0}$ and $\Sigma = I$. Partition **X**, μ, and Σ as

$$\mathbf{X} = \begin{bmatrix} \mathbf{X}_1 \\ \mathbf{X}_2 \end{bmatrix}, \qquad \mu = \begin{bmatrix} \mu_1 \\ \mu_2 \end{bmatrix}$$

$$\Sigma = \begin{bmatrix} \Sigma_{11} & \Sigma_{12} \\ \Sigma_{22} & \Sigma_{22} \end{bmatrix},$$

where $\mathbf{X}_1$ is $k \times 1$, μ_1 is $k \times 1$, Σ_{11} is $k \times k$, and so forth. The density

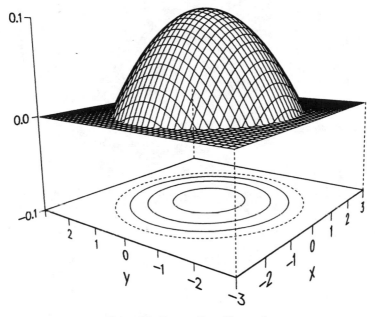

Figure 6.5. Pearson Type II, $m = 1$.

function of $\mathbf{X}_1$ given $\mathbf{X}_2 = \mathbf{x}_2$ is

$$f(\mathbf{x}_1 | \mathbf{X}_2 = \mathbf{x}_2) = \frac{\Gamma(m + k/2 + 1)}{\Gamma(m + 1)\pi^{k/2}} (1 - \mathbf{x}_2'\mathbf{x}_2)^{-k/2} \left[1 - \frac{\mathbf{x}_1'\mathbf{x}_1}{1 - \mathbf{x}_2'\mathbf{x}_2}\right]^m,$$

with support $\mathbf{x}_1'\mathbf{x}_1 \leqslant 1 - \mathbf{x}_2'\mathbf{x}_2$. This distribution is identical in form to (6.6) with the exception of the reduced support.

The results on marginal and conditional distributions for the spherical form of (6.6) could be exploited to derive an algorithm for generating the Pearson Type II multivariate distribution. The first component X_1 is generated as:

1. Generate X_1 having a beta $(\frac{1}{2}, m + p/2 + \frac{1}{2})$ distribution.
2. Set $X_1 = \pm X_1^{1/2}$, where "$\pm$" denotes a random sign (equally likely positive or negative).

After k components $X_1, X_2, \ldots, X_k$ have been generated, the $k + 1$ com-

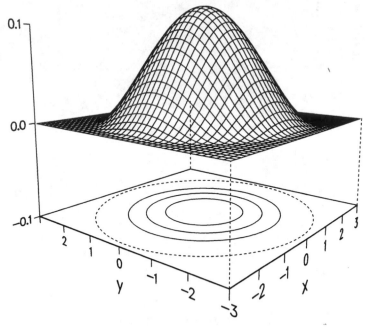

Figure 6.6. Pearson Type II, $m = 2$.

ponent is generated as:

1. Generate V_{k+1} having a beta $(\frac{1}{2}, m + p/2 + \frac{1}{2} - k/2)$ distribution.
2. Set $X_{k+1} = \pm[V_{k+1}(1 - X_1^2 - \cdots - X_k^2)]^{1/2}$.

The fully parameterized elliptical form (6.6) is then obtained as $\Sigma^{1/2}\mathbf{X} + \boldsymbol{\mu}$, where $\Sigma^{1/2}$ is a factorization of Σ. To save a few multiplications, the lower triangular Choleski factorization is recommended. For two and three dimensions, explicit forms of lower triangular $\Sigma^{1/2}$ were given in Section 4.1.

Alternatively, the Cambanis representation of (6.5) can be employed to generate the multivariate Pearson Type II. This scheme is particularly attractive here since $R^2 = (\mathbf{X} - \boldsymbol{\mu})'\Sigma^{-1}(\mathbf{X} - \boldsymbol{\mu})$ has a beta distribution with parameters $p/2$ and $m + 1$ (Section 2.1). Methods for generating the uniform distribution on the p-dimensional unit hypersphere are given later in Section 7.1. The Cambanis representation generates only one beta variate, so it is less cumbersome to implement than the algorithm induced by the conditional distributions, which uses a different beta variate for each component.

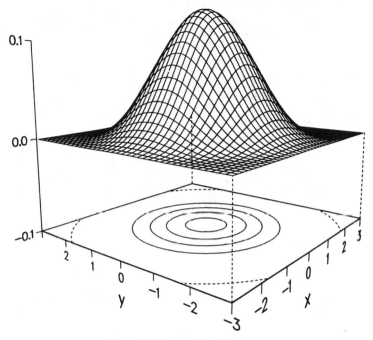

Figure 6.7. Pearson Type II, $m = 4$.

A natural way to compare the various Pearson Type II distributions is by consideration of the distribution of the multiplier variate R, which is the square root of a beta $(p/2, m + 1)$ random variable. Considerations in this manner can be helpful for selecting particular Pearson Type II distributions in a Monte Carlo study. To account for changes in scale of the components of $\mathbf{X}$ as m and p are altered, the diagonal matrix Σ is selected to have as nonzero entries the values $(2m + p + 2)$. This ensures that $\mathrm{Var}(X_i) = 1$ for any choice of m and p. Figure 6.8 gives density functions of $\sqrt{R}$ for various choices of m and p.

Pearson Type VII

The density function of the multivariate Pearson Type VII distribution is

$$f(\mathbf{x}) = \frac{\Gamma(m)}{\Gamma(m - p/2)\pi^{p/2}}|\Sigma|^{-1/2}\big[1 + (\mathbf{x} - \mu)'\Sigma^{-1}(\mathbf{x} - \mu)\big]^{-m}, \quad (6.8)$$

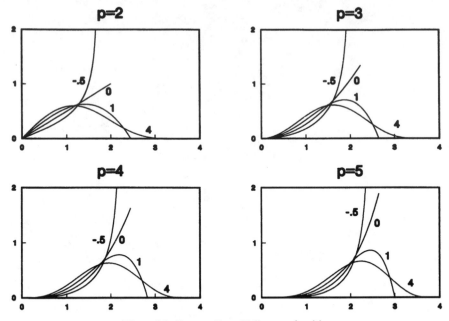

Figure 6.8. Pearson Type II distance densities.

where $m > p/2$ and $-\infty < x_i < \infty$, $i = 1, 2, \ldots, p$. An alternative parameterization is embodied in the density function (Johnson and Kotz, 1972, p. 134)

$$g(\mathbf{y}) = \frac{\Gamma[(\nu + p)/2]}{(\pi\nu)^{p/2}\Gamma(\nu/2)}|\Sigma|^{-1/2}\big[1 + \nu^{-1}(\mathbf{y} - \mu)'\Sigma^{-1}(\mathbf{y} - \mu)\big]^{-(\nu+p)/2},$$

with $\nu > 0$. This form is known as a general multivariate t distribution with ν degrees of freedom. The distribution can be obtained by the transformation

$$\mathbf{Y} = \left(\frac{\sqrt{S}}{\nu}\right)^{-1}\mathbf{Z} + \mu, \qquad (6.9)$$

where $\mathbf{Z}$ is $N_p(\mathbf{0}, \Sigma)$ and independent of S, which is $\chi^2_{(\nu)}$. The case with $\nu = 1$ is a multivariate Cauchy distribution.

The functional form (6.8) will be used to facilitate the description of distributional properties. The parameters μ and Σ do not necessarily admit a ready interpretation in terms of moments. The kth moment of X_i, the ith component of $\mathbf{X}$ in (6.8), exists only for $m > (p + k)/2$. If $m > (p + 1)/2$, then $E(\mathbf{X}) = \mu$. For all values of $m > p/2$, the permissible range of m, the vector μ is the mode of the density. If $m > (p + 2)/2$, then

$$\text{Cov}(\mathbf{X}) = \frac{1}{2m - p - 2}\Sigma.$$

Some bivariate density plots are provided in Figures 6.9–6.13. These plots use $\mu = 0$ and Σ the identity matrix. Unlike the Pearson Type II distribution, we cannot choose the entries of Σ to produce equal variances in the marginal distributions. In spite of varying scales, the approach to bivariate normality as m gets large is quite apparent.

Marginal and conditional distributions can be simply stated in terms of the density function (6.8). The marginal distribution of k components ($k < p$) of $\mathbf{X}$ is Pearson Type VII with shape parameter $m - [(p - k)/2]$ and location and scale parameters extracted from a partitioned μ and Σ. The conditional distributions are especially easy to give for the spherical form with $\Sigma = I$ and $\mu = 0$. If $\mathbf{X}$ is partitioned as $\mathbf{X}' = (\mathbf{X}_1', \mathbf{X}_2')$ with $\mathbf{X}_1$

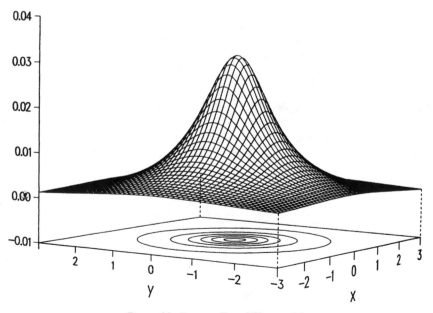

Figure 6.9. Pearson Type VII, $m = 1.1$.

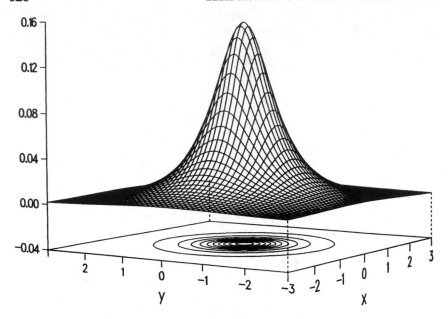

Figure 6.10. Cauchy marginals for Pearson Type VII, $m = 1.5$.

having dimensions $k \times 1$, then the conditional distribution of $\mathbf{X}_1$ given $\mathbf{X}_2 = \mathbf{x}_2$ has a k-dimensional Pearson Type VII distribution with shape parameter m and diagonal entries equal to a constant value $(1 + \mathbf{x}_2'\mathbf{x}_2)^{-1}$.

The expression given in (6.9) is immediately suitable for generating random vectors having a Pearson Type VII multivariate distribution. This approach requires a p-dimensional normal random vector and an independent $\chi^2_{(2m-p)}$ variate. Alternatively, a conditional distribution approach could be developed using the results just presented. A third algorithm is possible from the Cambanis representation. The distribution of $Z = (\mathbf{X} - \boldsymbol{\mu})'\Sigma^{-1}(\mathbf{X} - \boldsymbol{\mu})$ in (6.8) has density function

$$g(z) = \frac{\Gamma(m)}{\Gamma(p/2)\Gamma(m - p/2)} z^{p/2-1}(1 + z)^{-m}, \qquad z > 0, \quad (6.10)$$

which is a univariate Pearson Type VI. Variates having this density can be generated as

$$Z = \frac{Y}{1 - Y},$$

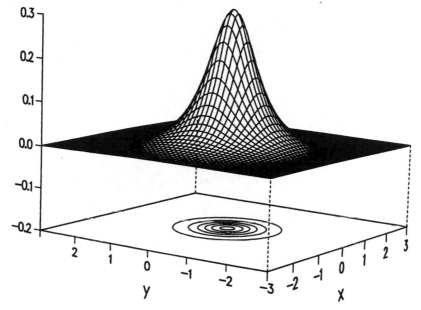

Figure 6.11. Pearson Type VII, $m = 2.0$.

where Y is beta $(p/2, m - p/2)$. The Cambanis approach then involves

$$\mathbf{X} = Z^{1/2}LU^{(p)} + \mu,$$

where $LL' = \Sigma$ and $U^{(p)}$ is uniform on the p-dimensional unit hypersphere. The preferred algorithm would appear to be the one based on (6.9), since it is so simple.

Other Elliptically Contoured Distributions

The Pearson Types II and VII and the multivariate normal distributions probably include an ample number of elliptically contoured distributions for most Monte Carlo studies. These distributions are not, however, the only elliptically contoured distributions that have been examined by researchers. The bivariate Laplace and generalized Laplace distributions (McGraw and Wagner, 1968 and Johnson and Kotz, 1972) can be mentioned. Their parametric forms involve Bessel functions, which inhibits their use by a broad spectrum of researchers. Moreover, variate generation has not been explicitly worked out for these distributions in the literature. Some

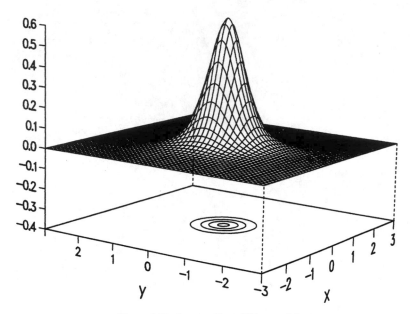

Figure 6.12. Pearson Type VII, $m = 3.0$.

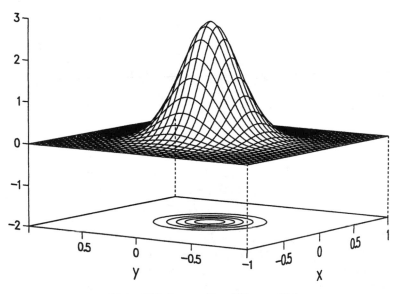

Figure 6.13. Pearson Type VII, $m = 10.0$.

innovative work on series distributions by Devroye (1981) could provide some techniques to accomplish it.

A point alluded to in Section 4.2 was that certain scale contaminated multivariate normal distributions are elliptically symmetric. For example, the distribution denoted $\alpha N_p(\mu, \Sigma) + (1 - \alpha)N_p(\mu, \beta\Sigma)$ is elliptically contoured, although it seems very limited in contrast to the general contaminated normal distributions pursued in Section 4.2.

Rather than examining the elliptically contoured class to exhaustion, a desirable strategy would be to consider other distributions that could help to explain results already obtained. For example, random vectors with

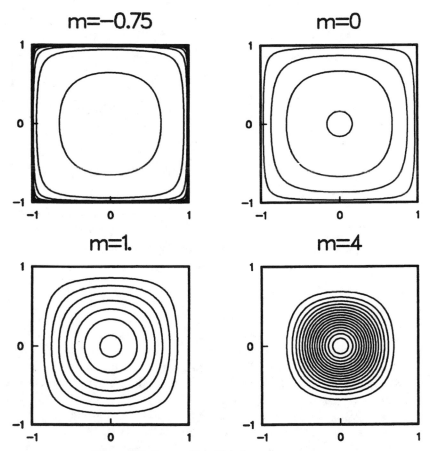

Figure 6.14. Pearson Type II, independent components.

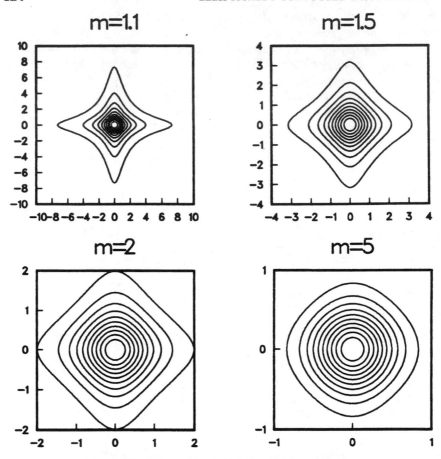

Figure 6.15. Pearson Type VII, independent components.

independent components having the same component distributions as the
marginal distributions of certain Pearson Types II and VII distributions
could shed some light on the effects of the dependence structure. In the case
of the Pearson Type II marginal distributions, the independent components
approach leads to distributions having different supports than those found
in (6.6). Figures 6.14 and 6.15 contrast the distributions with independent
components versus the analogous bivariate Pearson Types II and VII
distributions with the same marginal distributions. If the corresponding
moments exist, the Pearson distributions are uncorrelated but they do not
have independent components. Differences in Monte Carlo results from
such pairs of distributions could be rationally attributed to the dependence
structure of the Pearson distributions.

CHAPTER 7

Circular, Spherical, and Related Distributions

This chapter considers a diverse set of distributions. Their commonality rests with distributions restricted to the circle, sphere, or the n-dimensional unit hypersphere (n-sphere for short). Of these we first consider the simple uniform distribution (Section 7.1), for which a variety of generation algorithms is described. Nonuniform distributions are considered in Section 7.2. The treatment of the classical circular and spherical forms focuses on variate generation since thorough mathematical coverage of these distributions is available in Watson (1983) and Mardia (1972). The emphasis in Section 7.2 is on the development of some new distributions that have particular appeal to robustness studies.

The material presented in this chapter is based on an impending paper (Nachtsheim and Johnson, 1986) that uses these distributions in a simulation study of Hotelling's T^2, as suggested in Section 1.3.

7.1. UNIFORM DISTRIBUTIONS

The uniform distribution on the n-sphere corresponds to equally likely directions. The n-sphere is defined as the set of points $\mathbf{x} \in R^n$ for which $\mathbf{x}'\mathbf{x} = 1$. A point $\mathbf{x}$ on the n-sphere can be uniquely represented by $n - 1$

angles $\theta_1, \theta_2, \ldots, \theta_{n-1}$ and the equations:

$$x_1 = \sin\theta_1 \sin\theta_2 \ldots \sin\theta_{n-2}\sin\theta_{n-1}$$

$$x_2 = \sin\theta_1 \sin\theta_2 \ldots \sin\theta_{n-2}\cos\theta_{n-1}$$

$$x_3 = \sin\theta_1 \sin\theta_2 \ldots \sin\theta_{n-3}\cos\theta_{n-2}$$

$$\vdots$$

$$x_{n-2} = \sin\theta_1 \sin\theta_2 \cos\theta_3$$

$$x_{n-1} = \sin\theta_1 \cos\theta_2$$

$$x_n = \cos\theta_1. \tag{7.1}$$

The uniform distribution on the n-sphere results if the angles $(\Theta_1, \ldots, \Theta_{n-1})$ have the distribution with density function

$$g(\theta_1, \ldots, \theta_{n-1}) \propto \sin^{n-2}\theta_1 \sin^{n-3}\theta_2 \ldots \sin\theta_{n-2},$$

$$0 < \theta_i < \pi, \ i = 1, \ldots, n-2; \ 0 < \theta_{n-1} < 2\pi. \tag{7.2}$$

From this functional form, it is apparent that the random angles $\Theta_1, \ldots, \Theta_{n-1}$ are independent random variables. The variate Θ_{n-1} is uniform on $(0, 2\pi)$, while the other angles have power sine densities. Recall that in Section 2.1, a direct method was given for generating Θ_{n-2} (since $g(\theta_{n-2}) \propto \sin\theta_{n-2}$) and a rejection method was devised for the other power sine distributions. Hence, in principle, we have an algorithm for generating the uniform distribution on the n-sphere:

1. Generate Θ_{n-1} as $2\pi U$, where U is uniform 0–1.
2. Generate Θ_{n-2} as $\cos^{-1}(1 - 2V)$, where V is uniform 0–1.
3. Generate $\Theta_1, \ldots, \Theta_{n-3}$ using the rejection method in Section 2.1.
4. Evaluate x according to (7.1).

For the uniform distribution on the circle, this generation scheme simplifies to

$$U_1 = \cos(2\pi U)$$

$$U_2 = \sin(2\pi U), \tag{7.3}$$

where U is uniform 0–1. Note that the right-hand side of this equation is part of the Box-Muller transformation (equation 2.5). Aside from two dimensions, the polar method embodied in (7.1) and (7.2) has not been used for generation purposes because of the far more convenient method described next.

The usual method for generating uniform variates on the n-sphere is to appeal to the distribution's relation with the multivariate normal distribution (Muller, 1959). If $X_1, X_2, \ldots, X_n$ are independent standard normal variates, the vector $\mathbf{U}$ determined by

$$U_i = \frac{X_i}{\left(X_1^2 + \cdots + X_n^2 \right)^{1/2}}, \qquad i = 1, 2, \ldots, n, \qquad (7.4)$$

has the uniform distribution (7.1). The obvious advantages of (7.4) include simplicity, the availability of normal generators, and the applicability to arbitrary dimensions. The method given by (7.4) is to be preferred from the generation efficiency standpoint. However, the polar approach embodied in (7.1) and (7.2) lends itself to the construction of some interesting new distributions (Section 7.2).

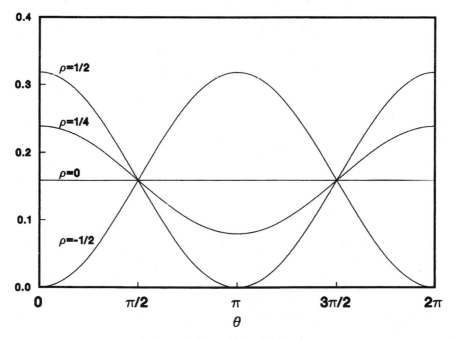

Figure 7.1. Cardioid density functions.

The method (7.4) is insufficiently fast to please various researchers, who not surprisingly have developed efficient alternatives. Cook's (1957) method for three dimensions is quite simple. Generate independent uniform variates on $(-1, 1)$, say U_1, U_2, U_3, and U_4, until the condition $S = U_1^2 + U_2^2 + U_3^2 + U_4^2 < 1$. The desired variate is

$$\left[\frac{2(U_1U_3 + U_2U_4)}{S}, \frac{2(U_3U_4 - U_1U_2)}{S}, \frac{U_1^2 + U_4^2 - U_2^2 - U_3^2}{S} \right].$$

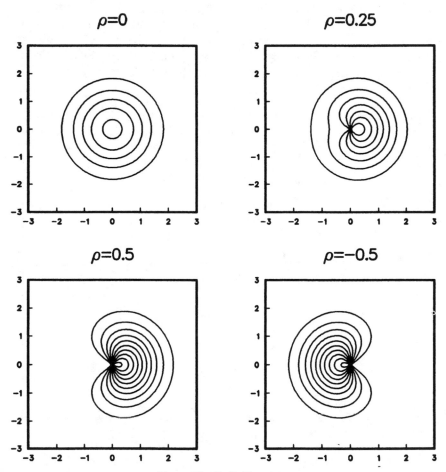

Figure 7.2. Cardioid contours.

Marsaglia's (1972) method is faster. Generate U_1 and U_2 independent uniform variates on $(-1, 1)$ until the condition $S = U_1^2 + U_2^2 < 1$ is satisfied. The desired variate is

$$\left[2U_1(1 - S)^{1/2}, 2U_2(1 - S)^{1/2}, 1 - 2S\right].$$

Marsaglia developed a similar method for four dimensions. Generate independent uniform variates U_1 and U_2 such that $S_1 = U_1^2 + U_2^2 < 1$. Also, generate independent uniform variates U_3 and U_4 such that $S_2 = U_3^2 + U_4^2 < 1$. The point on the 4-sphere,

$$\left[U_1, U_2, U_3 \left(\frac{1 - S_1}{S_2}\right)^{1/2}, U_4 \left(\frac{1 - S_1}{S_2}\right)^{1/2}\right],$$

has a uniform distribution.

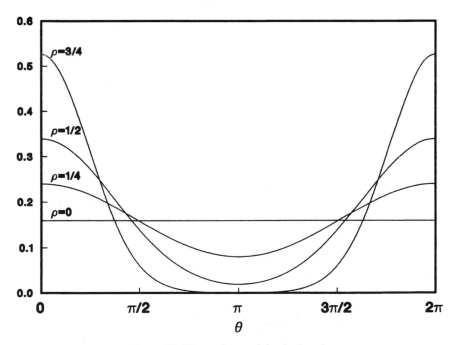

Figure 7.3. Wrapped normal density functions.

The above algorithms are restricted to three or four dimensions (although Marsaglia notes that his ideas can be applied in n dimensions). Hicks and Wheeling (1959) proposed the following algorithm:

1. Generate U_1 uniform 0–1. If $U_1 < 0.5$, set $x_1 = 1$. Otherwise set $x_1 = -1$.
2. Set $m = 2$.
3. Generate a uniform variate U_m on $(-1, 1)$.

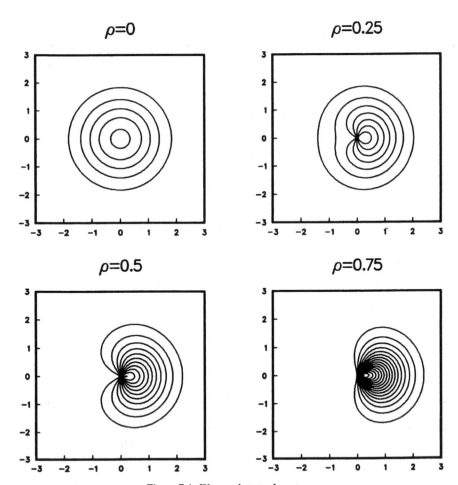

Figure 7.4. Wrapped normal contours.

4. Solve $F(r_m)/F(1) = |U_m|$ for r_m where $F(x) = \int_0^x \rho^{m-2}/\sqrt{1-\rho^2}\,d\rho$.

5. Form the m-dimensional point $(r_m x_1, r_m x_2, \ldots, r_m x_{m-1}, S\sqrt{1-r_m^2})$, where $S = u_m/|u_m|$.

6. If $m = n$, the point in step 5 has a uniform distribution on the n-sphere.

7. If $m < n$, set $m = m + 1$ and return to step 3.

Obviously, step 4 is a bit inconvenient since it involves multiple evaluations (after transformation anyway) of the incomplete beta function.

A more direct n-dimensional method has been given by Tashiro (1977), whose analysis of the method is evidently simpler than Sibuya's (1962). The algorithm takes one of two forms, depending on whether the dimension is even or odd. For notation, we let $X_1, X_2, \ldots$ and $\Theta_1, \Theta_2, \ldots$ be independent uniform 0–1 variates.

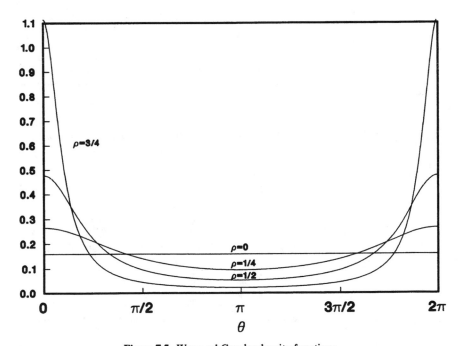

Figure 7.5. Wrapped Cauchy density functions.

Even Dimensions ($n = 2p$)

1. Set $Y_p = 1$ and $Y_0 = 0$.
2. Recursively compute

$$Y_{p-1} = Y_p X_{p-1}^{1/(p-1)}$$
$$Y_{p-2} = Y_{p-1} X_{p-2}^{1/(p-2)}$$
$$\vdots$$
$$Y_1 = Y_2 X_1.$$

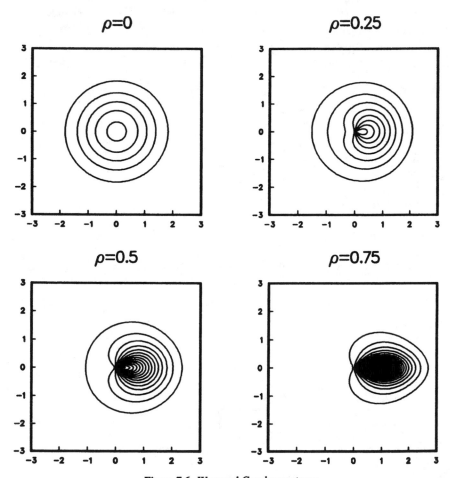

Figure 7.6. Wrapped Cauchy contours.

3. Now evaluate
$$y_{2i-1} = \sqrt{Y_i - Y_{i-1}} \cos 2\pi\Theta_i, \qquad i = 1, 2, \ldots, p$$
$$y_{2i} = \sqrt{Y_i - Y_{i-1}} \sin 2\pi\Theta_i, \qquad i = 1, 2, \ldots, p.$$

The desired variate on the n-sphere has components $y_1, \ldots, y_n$.

Odd Dimensions ($n = 2p + 1$)

1. Set $Z_{p+1} = 1$.
2. Recursively compute
$$Z_p = Z_{p+1} X_p^{2/(2p-1)}$$
$$Z_{p-1} = Z_p X_{p-1}^{2/(2p-3)}$$
$$\vdots$$
$$Z_1 = Z_2 X_1^2.$$

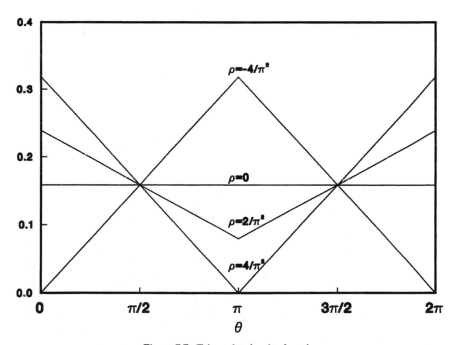

Figure 7.7. Triangular density functions.

3. Now evaluate

$$y_1 = \pm \sqrt{Z_1}$$

$$y_{2i-1} = \sqrt{Z_i - Z_{i-1}} \cos 2\pi\Theta_i, \qquad i = 1, 2, \ldots, p$$

$$y_{2i} = \sqrt{Z_i - Z_{i-1}} \sin 2\pi\Theta_i, \qquad i = 1, 2, \ldots, p.$$

The symbol "$\pm$" indicates that a random sign is to be attached to $\sqrt{Z_1}$. The desired variate on the n-sphere has components $y_1, \ldots, y_n$.

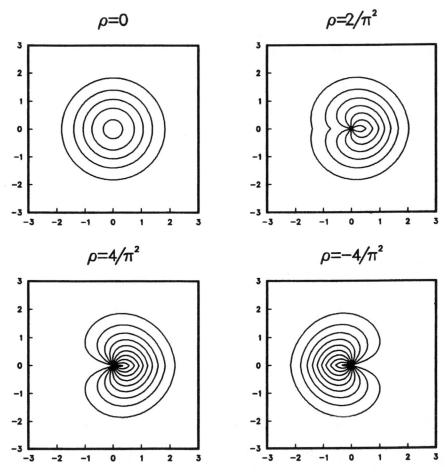

Figure 7.8. Triangular contours.

The above methods may be of interest from a mathematical and computationally efficient standpoint. However, the moderate gains in execution speed are often offset by foul-ups in implementing the more complicated algorithms. Hence, if uniform variates on the n-sphere are to be generated, then Muller's method of projecting independent normal variates onto the sphere is recommended.

7.2. NONUNIFORM DISTRIBUTIONS

Circular Case

The distributions examined first in this section rely on the polar form construction:

$$X = R \cos \Theta$$

$$Y = R \sin \Theta, \qquad (7.5)$$

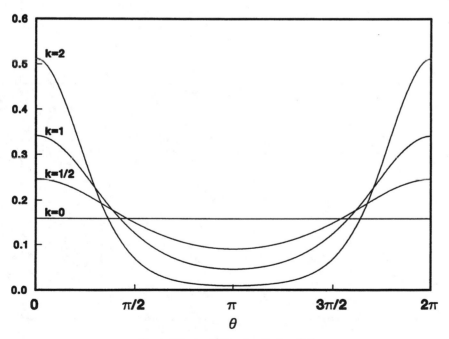

Figure 7.9. von Mises density functions.

where $0 < \Theta < 2\pi$ and $R > 0$. A variety of distributions (R, Θ) are posed and then the resulting distribution (X, Y) is examined. If the distribution of (Θ, R) has density function $g(\theta, r)$, then the density function of (X, Y) is easy to derive. The inverse of the polar transformation is

$$r = \left(x^2 + y^2\right)^{1/2}$$

$$\theta = \tan^{-1}(y/x),$$

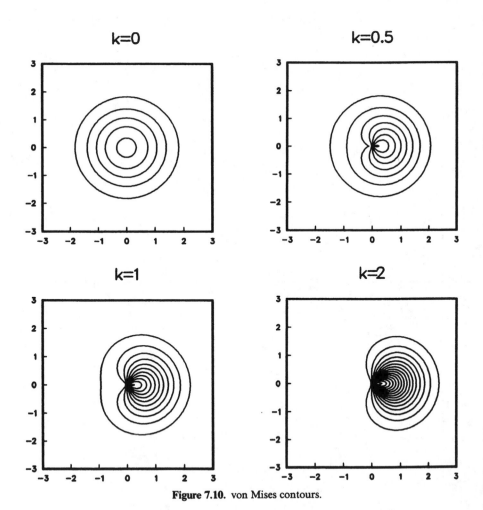

Figure 7.10. von Mises contours.

and the Jacobian is

$$\det|J| = \begin{vmatrix} \dfrac{x}{(x^2 + y^2)^{1/2}} & \dfrac{y}{(x^2 + y^2)^{1/2}} \\[2ex] \dfrac{-y/x^2}{1 + y^2/x^2} & \dfrac{1/x}{1 + y^2/x^2} \end{vmatrix}$$

$$= (x^2 + y^2)^{-1/2}.$$

Thus the density function of (X, Y) is

$$f(x, y) = (x^2 + y^2)^{-1/2} g\left[\tan^{-1}(y/x), (x^2 + y^2)^{1/2}\right]. \quad (7.6)$$

To facilitate comparisons with the usual bivariate normal distribution, we can further assume that R is distributed $\sqrt{\chi^2_{(2)}}$ and is independent of Θ.

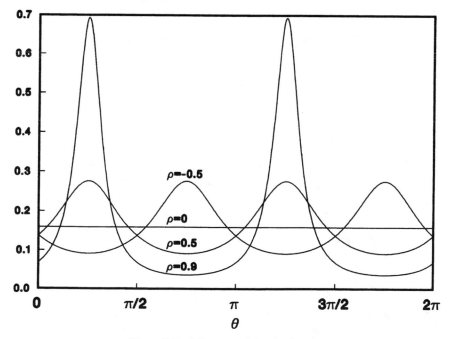

Figure 7.11. Offset normal density functions.

The density function of (R, Θ) is then of the form

$$g(r, \theta) = r \exp(-r^2/2) h(\theta),$$

implying a density of (X, Y) as

$$f(x, y) = \exp\left[\frac{-(x^2 + y^2)}{2}\right] h\left[\tan^{-1}(y/x)\right]. \tag{7.7}$$

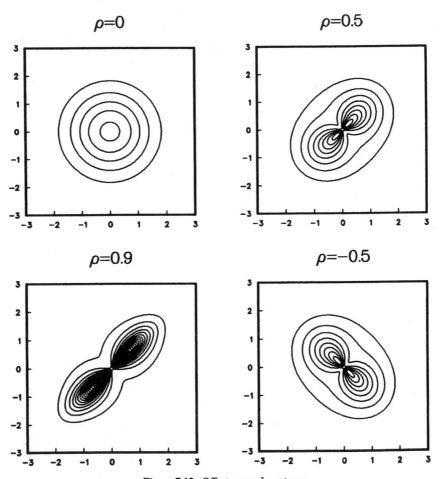

Figure 7.12. Offset normal contours.

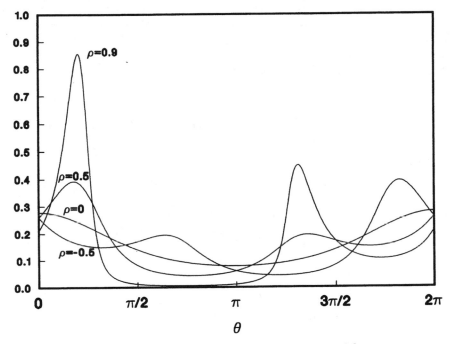

Figure 7.13. Offset normal density functions, $\mu_1 = 0.5$.

It is an easy matter now to pose candidate circular distributions for h and assess their impact on f. Table 7.1, adapted from Nachtsheim and Johnson (1986), lists some classical circular densities h and the corresponding f.

Figures 7.1–7.20 provide density plots of $h(\theta)$ and the corresponding densities of (X, Y) given by (7.7). For θ values having $h(\theta)$ near zero, little mass for f can be found in the direction $(\cos\theta, \sin\theta)$. Similarities across the various cases are quite apparent. The cardioid, wrapped normal, and von Mises distributions agree well both in contours and in the function $g(\theta)$. The wrapped Cauchy has similar contour shapes, but under the present parameterization the contours are more congested. The triangular distribution is at best a crude approximation to the cardioid distribution, but produces contour plots for $f(x, y)$ that match up fairly well. The offset normal distribution seems to have the greatest facility for yielding rather different looking contour plots. However, for appropriate parameter choices in this distribution, agreement with the other contour plots can be obtained.

Few analytical properties of the (X, Y) distributions have been worked out, although this effort is underway. Many features of interest for these

TABLE 7.1. Two-Dimensional Distributions

General
$h(\theta)$: $\quad h(\theta)$

$f(x, y)$: $\quad h[\tan^{-1}(y/x)]\exp[-(x^2 + y^2)/2]$

Cardioid
$h(\theta)$: $\quad (1 + 2\rho \cos\theta)/(2\pi)$

$f(x, y)$: $\quad \exp[-(x^2 + y^2)/2][1 + 2\rho x/(x^2 + y^2)^{1/2}]\big/(2\pi)$

Offset Normal
$h(\theta)$: $\quad [C(\theta)]^{-1}[\phi_2(\mu_1, \mu_2; \mathbf{0}, \Sigma) + aD(\theta)\Phi[D(\theta)]$
$\qquad\qquad \times \phi_1\{a[C(\theta)]^{-1/2}(\mu_1\sin\theta - \mu_2\cos\theta)\}],$

where
$a = [\sigma_1\sigma_2(1 - \rho^2)^{1/2}]^{-1}$
$C(\theta) = a^2(\sigma_2^2\cos^2\theta - \rho\sigma_1\sigma_2\sin 2\theta + \sigma_1^2\sin^2\theta)$
$D(\theta) = a^2[C(\theta)]^{-1/2}[\mu_1\sigma_2(\sigma_2\cos\theta - \rho\sigma_1\sin\theta)$
$\qquad + \mu_2\sigma_1(\sigma_1\sin\theta - \rho\sigma_2\cos\theta)]$
ϕ_1 is density of $N(0, 1)$
ϕ_2 is density of $N_2(\mathbf{0}, \Sigma)$
Φ is distribution function of $N(0, 1)$

$f(x, y)$: $\quad$ substitute using the general distribution

Triangular
$h(\theta)$: $\quad (8\pi)^{-1}[4 - \pi^2\rho + 2\pi\rho|\pi - \theta|]$

$f(x, y)$: $\quad \exp[-(x^2 + y^2)/2](8\pi)^{-1}[4 - \pi^2\rho + 2\pi\rho|\pi - \tan^{-1}(y/x)|]$

Wrapped Cauchy

$h(\theta)$: $\quad \dfrac{1 - \rho^2}{2\pi(1 + \rho^2 - 2\rho \cos\theta)}$

$f(x, y)$: $\quad \dfrac{(1 - \rho^2)\exp[-(x^2 + y^2)/2]}{2\pi[1 + \rho^2 - 2\rho x(x^2 + y^2)^{-1/2}]}$

Wrapped Normal
$h(\theta)$: $\quad (2\pi\sigma^2)^{-1/2}\sum_{k=-\infty}^{\infty}\exp[-(\theta + 2\pi k)^2/(2\sigma^2)]$

$f(x, y)$: $\quad$ substitute using the general distribution

von Mises
$h(\theta)$: $\quad c \exp(k \cos\theta)$

$f(x, y)$: $\quad c \exp[kx/(x^2 + y^2)^{1/2} - (x^2 + y^2)/2]$

Power Sine
$h(\theta)$: $\quad c \sin^k(\theta/2)$

$f(x, y)$: $\quad \exp[-(x^2 + y^2)/2][y/(x^2 + y^2)^{1/2}]^k$

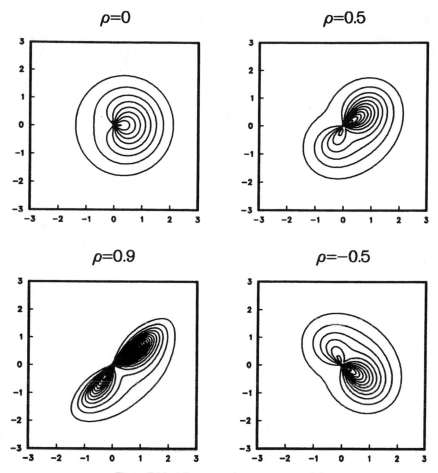

Figure 7.14. Offset normal contours, $\mu_1 = 0.5$.

distributions can be estimated from Monte Carlo exercises. To accomplish such calculations, the following algorithms are offered for generating the classical circular distributions.

Cardioid

The density function of the cardioid assumes its maximum at θ equal to θ_0 or $\theta_0 + \pi$ depending on the sign of ρ. The following rejection algorithm

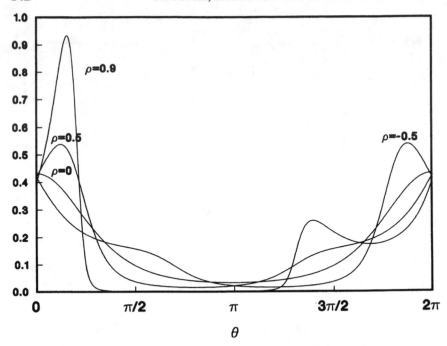

Figure 7.15. Offset normal density functions, $\mu_1 = 1$.

applies:

1. Generate U_1, U_2 independent uniform 0–1 variates.
2. If $U_2 \leqslant [1 + 2\rho \cos(2\pi U_1 - \theta_0)]/(1 + 2|\rho|)$, accept $\Theta = 2\pi U_1$. Otherwise, return to step 1.

Triangular

As with the cardioid distribution, a rejection method is easily developed since the density function has an obvious maximum:

1. Generate U_1, U_2 independent uniform 0–1 variates.
2. If $U_2 \leqslant (4 - \pi^2\rho + 2\pi\rho|\pi - \theta|)/(4 + |\rho|\pi^2)$, accept $\Theta = 2\pi U_1$. Otherwise, return to step 1.

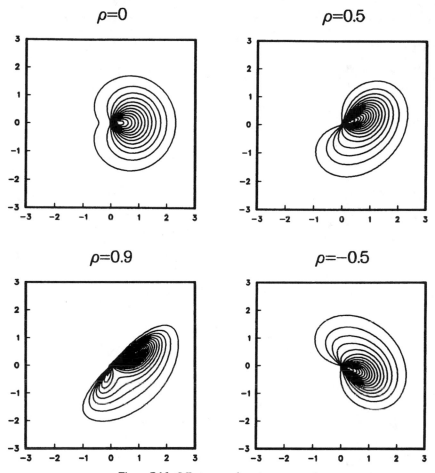

Figure 7.16. Offset normal contours, $\mu_1 = 1$.

Offset Normal

The offset normal distribution enjoys a simple generation scheme by virtue of its genesis. The distribution arises by considering the positive angle formed by the x-axis and a realization of a bivariate normal distribution with parameters μ_1, σ_1, μ_2, σ_2, and ρ. The following algorithm is offered:

1. Generate Y_1, Y_2 independent, standard normal.
2. Compute:
   $$X_1 = \sigma_1 Y_1 + \mu_1$$
   $$X_2 = \sigma_2[\rho Y_1 + \sqrt{1 - \rho^2}\, Y_2] + \mu_2$$

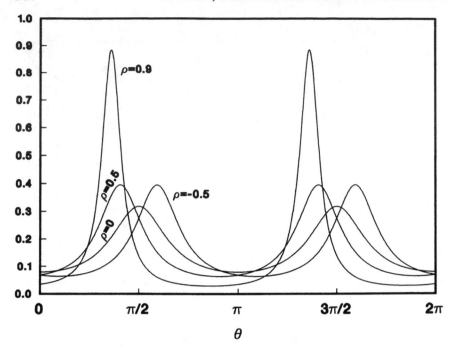

Figure 7.17. Offset normal density functions, $\sigma_1 = 0.5$.

$$W = \frac{X_1}{\sqrt{X_1^2 + X_2^2}}.$$

3. Set $\theta = \cos^{-1}(W)$ $X_2 > 0$

       $= 2\pi - \cos^{-1}(W)$ $X_2 < 0$

The two cases noted in step 3 are necessary to ensure that $\Theta \in (0, 2\pi)$.

Wrapped Normal and Cauchy

If X is standard normal, then $\Theta = (\sigma X) \bmod(2\pi)$ yields the desired wrapped normal distribution. Similarly, for the wrapped Cauchy, take $\Theta = \tan[\pi(U - \frac{1}{2})] \bmod(2\pi)$, where U is uniform 0–1.

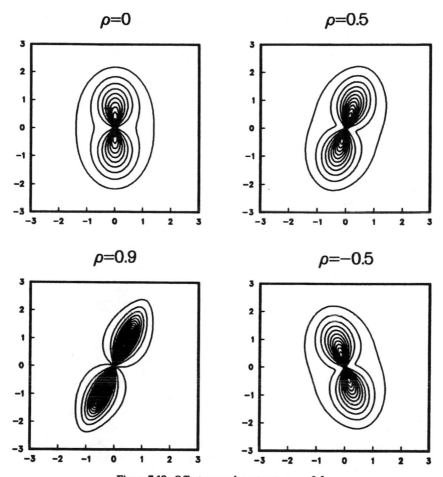

Figure 7.18. Offset normal contours, $\sigma_1 = 0.5$.

von Mises

An easy rejection method is available since the maximum of $\exp(k \cos \theta)$ is $\exp(k)$. Hence:

1. Generate U_1, U_2 independent uniform 0–1 variates.
2. If $U_2 \leq \exp[k(\cos(2\pi U_1 - 1))]$, accept $\Theta = 2\pi U_1$. Otherwise, return to step 1.

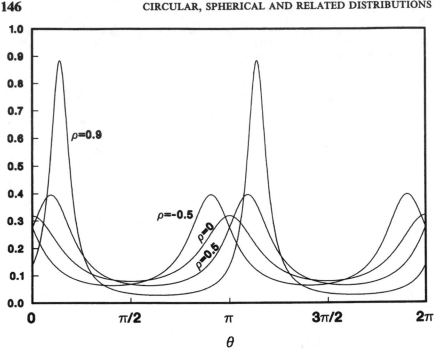

Figure 7.19. Offset normal density functions, $\sigma_1 = 2$.

Spherical Case

For spherical distributions, both the classical Fisher and Bingham distributions were mentioned in earlier chapters. In Section 2.1, the difficult part of generating Fisher's distribution was overcome. In particular, an algorithm was given for generating an angle Θ on the interval $(0, \pi)$. The other angle Φ in Fisher's distribution is uniform on $(0, 2\pi)$. The resulting joint density function of (Θ, Φ) is

$$g(\theta, \phi) = \frac{\kappa}{\sqrt{4\pi} \sinh(\kappa)} \exp(\kappa \cos \theta) \sin \theta, \qquad 0 < \theta < \pi, 0 < \phi < 2\pi.$$

Bingham's distribution was introduced in Section 3.1 and a complete generation algorithm was provided. In the same spirit as in the circular

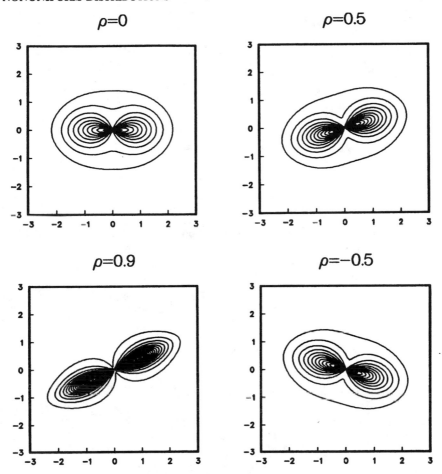

Figure 7.20. Offset normal contours, $\sigma_1 = 2$.

case, "new" distributions arise using the construction

$$X = R \sin\theta \sin\phi$$

$$Y = R \sin\theta \cos\phi$$

$$Z = R \cos\theta,$$

where (Θ, Φ) has a spherical distribution with density function g and R^2 has an independent χ^2 distribution with three degrees of freedom. The resulting density function of (X, Y, Z) is

$$f(x, y, z) = \sqrt{\frac{2}{\pi}} \, \exp\left[\frac{-(x^2 + y^2 + z^2)}{2}\right] g(\theta, \phi),$$

where g is evaluated at

$$\theta = \cos^{-1}\left(\frac{z}{x^2 + y^2 + z^2}\right)$$

$$\phi = \tan^{-1}\left(\frac{y}{x}\right).$$

For both the three-dimensional Bingham and Fisher spherical distributions, considerable simplification takes place in the form of f. For Bingham's distribution, the density in R^3 is

$$f(x, y, z) = c\sqrt{\left(\frac{2}{\pi}\right)} \, \exp\left[\frac{-(x^2 + y^2 + z^2)}{2}\right] \exp\left[-\frac{\kappa_1 x^2 + \kappa_2 y^2 + \kappa_3 z^2}{x^2 + y^2 + z^2}\right].$$

For Fisher's distribution, the density in R^3 is

$$g(x, y, z) = \left(\frac{2}{\pi}\right)^{-3/2}\left[\frac{k}{\sinh(k)}\right]\exp\left[\frac{-(x^2 + y^2 + z^2)}{2}\right]$$

$$\times \exp\left(\frac{kz}{\sqrt{x^2 + y^2 + z^2}}\right).$$

General Case

The n-dimensional setting can be developed similarly to the above cases, using the construction (7.1) in concert with distributions for $(\Theta_1, \Theta_2, \ldots, \Theta_n)$ deviating from the uniform case (7.2).

CHAPTER 8

Khintchine Distributions

Khintchine's (1938) theorem on unimodality of univariate distributions can be exploited to construct a diverse set of multivariate distributions. The resulting distributions are easy to generate, can have a wide variety of dependence structures, and offer intriguing alternatives to the multivariate normal distribution (Section 4.1). The material of this chapter draws on a recent paper by Bryson and Johnson (1982). The theorem underlying these developments is first described.

8.1. KHINTCHINE'S UNIMODALITY THEOREM

A continuous random variable X has a single mode at zero if and only if it can be expressed as the product $X = ZU$ where Z and U are independent and U has the uniform 0–1 distribution.

Feller (1971, pp. 157–159) gives a proof of this theorem. If X has the differentiable density function f_X, it is easily verified by a change of variable argument that Z must have the density function

$$f_Z(z) = -zf_X'(z). \tag{8.1}$$

To see that the construction ZU leads to a unimodal density, Figure 8.1 provides a heuristic justification (by example). Suppose we condition on the five values of Z generated, denoted by the order statistics $z_{(1)}, \ldots, z_{(5)}$. The possible values of ZU given $z_{(1)}, \ldots, z_{(5)}$ are represented in the lower half of the figure. All realizations of Z contribute some mass near the origin to the distribution ZU.

Some applications of Khintchine's theorem were noted in Section 2.4. If X is standard normal, then using (8.1), Z has the distribution of $\sqrt{\chi_{(3)}^2}$ with

149

DENSITY FUNCTION OF Z

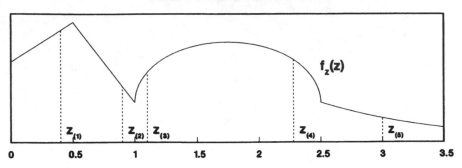

CONTRIBUTIONS TO DENSITY FUNCTION OF X

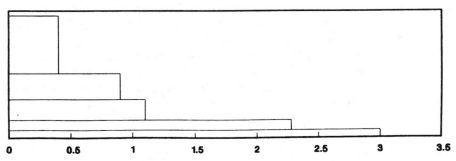

Figure 8.1. Khintchine's unimodality theorem. The upper figure gives a hypothetical density function f_z with five realizations. The lower plot indicates the contribution to the density function of X.

a random sign. More generally, if Z is a gamma variate raised to a power and subsequently a random sign is attached, then $Z \cdot U$ has the generalized exponential power distribution described in Section 2.4. As another example, if X is to be exponential, $\Gamma(1, \beta)$, then from (8.1) Z must be $\Gamma(2, \beta)$. The Khintchine representation $X = ZU$ is also quite useful for determining properties of unimodal distributions. In particular, it is easy to show that the coefficient of variation C_X (ratio of standard deviation to the mean of X) is greater than or equal to $3^{-1/2}$ for any unimodal distribution. We have

$$C_X = \frac{\left[E(X^2) - E^2(X)\right]^{1/2}}{E(X)},$$

and we know $E(X) = (\frac{1}{2})E(Z)$ and $E(X^2) = (\frac{1}{3})E(Z^2)$. The variance of

Z is non-negative so that

$$E(Z^2) - E^2(Z) \geq 0,$$

which in terms of X is

$$3E(X^2) - 4E^2(X) \geq 0.$$

This can be rewritten as

$$E(X^2) \geq \tfrac{4}{3}E^2(X)$$

$$E(X^2) - E^2(X) \geq \tfrac{1}{3}E^2(X)$$

$$C_X \geq \frac{1}{\sqrt{3}}.$$

As another result for a random variable X with a unique mode at a, the random variable $Y = X - a$ is unimodal at zero. Thus $C_Y \geq 3^{-1/2}$, which can be written in terms of X as

$$\frac{\text{mean of } X - \text{mode of } X}{\text{standard deviation of } X} \leq \sqrt{3}.$$

The left-hand side is known as Pearson's coefficient of skewness. This inequality was first proven by Johnson and Rogers (1951).

To exploit Khintchine's theorem in a multivariate framework, consider the following setup:

$$X_1 = Z_1 U_1$$

$$X_2 = Z_2 U_2$$

$$\vdots \tag{8.2}$$

$$X_p = Z_p U_p.$$

If each U_i is uniform 0–1 and independent of its paired Z_i, then each X_i has a unimodal distribution. The distribution of the individual Z_i's could be determined by the distribution desired for X_i. In some cases, multivariate versions of the distribution of Z_i can be used to build a dependence structure in the X_i's. Also, the set of U_i's can be conceived as a multivariate distribution with uniform marginals. Before leaping to the most general case with both Z and U having nonindependent multivariate distributions, it is worth considering simpler special cases to assess the utility of the approach in (8.2). Complicated forms for Z and U can lead to messy forms for X.

One approach in considering (8.2) is to model the distribution (U_1, U_2) with one of the bivariate uniform distributions. The possibilities include Morgenstern (Section 10.1), Plackett (Section 10.2), Gumbel (Section 10.3), Ali-Mikhail-Haq (Section 10.4), normal-uniform (i.e., apply the probability integral transform Φ to the components of a bivariate normal), and Burr-Pareto-logistic (Chapter 9) distributions. The random variables Z_1 and Z_2 will be termed the "generator" variables since they determine the marginal distributions of X_1 and X_2. The first set of results is derived from two simple assumptions on Z_1 and Z_2—identical generators $(Z_1 = Z_2)$ or independent generators with the same distribution. For reference, the bivariate form of (8.2) is given:

$$X_1 = Z_1 U_1$$

$$X_2 = Z_2 U_2, \tag{8.3}$$

with Z_i and U_i independent, $i = 1, 2$.

8.2. IDENTICAL GENERATORS

With $Z_1 = Z_2$, the density of (X_1, X_2) in (8.3) is

$$f(x_1, x_2) = \int_{-\infty}^{\infty} \frac{f_X'(w)}{w} g\left(\frac{x_1}{w}, \frac{x_2}{w}\right) dw, \tag{8.4}$$

where g is the density function of (U_1, U_2) and f_X is the marginal density common for X_1 and X_2. To assess roughly the range of dependence in (8.4), the correlation of (X_1, X_2) can be computed directly from (8.3). This yields

$$\rho(X_1, X_2) = \tfrac{1}{4}\left[3 - \frac{1}{C_X^2} + \rho(U_1, U_2)\left(1 + \frac{1}{C_X^2}\right)\right], \tag{8.5}$$

with $\rho(U_1, U_2)$ the correlation of (U_1, U_2) and C_X the common coefficient of variation of X_1 and X_2. If the X_i's have mean zero, the expression becomes

$$\rho(X_1, X_2) = \frac{3 + \rho(U_1, U_2)}{4}$$

with range $\frac{1}{2}$ to 1. For exponential marginals, the expression (8.5) is

$$\rho(X_1, X_2) = \frac{1 + \rho(U_1, U_2)}{2}$$

with range 0–1.

8.3. INDEPENDENT GENERATORS

Assume now that Z_1 and Z_2 are independent with identical distributions, so that X_1 and X_2 are identically distributed. The joint distribution of (X_1, X_2) has density

$$f(x_1, x_2) = \int_{-\infty}^{\infty} \int_{-\infty}^{\infty} f_X'(w_1) f_X'(w_2) g\left(\frac{x_1}{w_1}, \frac{x_2}{w_2}\right) dw_1 \, dw_2, \quad (8.6)$$

where again g denotes the density of (U_1, U_2) and f_x is the density of each component X_1 and X_2. Correlations can be computed directly from (8.3) under the independent generator assumption, thus circumventing manipulations to (8.6). Taking the appropriate expectations in (8.3) leads to

$$\rho(X_1, X_2) = \rho(U_1, U_2)(3C_X^2)^{-1},$$

where $\rho(U_1, U_2)$ is the correlation of (U_1, U_2) and C_X is the common coefficient of variation of X_1 and X_2. Since the X_i's are unimodal, we have that $C_X \geqslant 1/\sqrt{3}$ and $0 \leqslant (3C_X^2)^{-1} \leqslant 1$. Hence, the range of correlations depends on the distribution of the X_i's. If X_1 and X_2 have normal or other symmetric distributions with finite variances, then X_1 and X_2 are uncorrelated although they are independent only if U_1 and U_2 are. For exponential marginals the correlation range is $-\frac{1}{3}$ to $\frac{1}{3}$, which corresponds to weak dependence.

8.4. OTHER POSSIBILITIES

The density forms in (8.4) and (8.6) become quite tractable if we analo-
gously assume the same two possibilities for the uniform variates, that is,
independent or identical. The combinations and the resulting densities can
be conveniently summarized, as follows:

1. Independent uniforms, independent generators:
   $$f_1(x_1, x_2) = f_X(x_1)f_X(x_2).$$
2. Identical uniforms, independent generators:
   $$f_2(x_1, x_2) = \int_0^1 w^{-2} f_Z\left(\frac{x_1}{w}\right) f_Z\left(\frac{x_1}{w}\right) dw.$$
3. Independent uniforms, identical generators:
   $$f_3(x_1, x_2) = \int_{\max(|x_1|, |x_2|)}^{\infty} w^{-1} f_X'(w) \, dw.$$
4. Identical uniforms, identical generators:
   degenerate distribution, $X_1 = X_2.$

The integrals for cases 2 and 3 have pleasant forms for exponential and
normal X_i's. For exponential marginals,

$$f_2(x_1, x_2) = \frac{x_1 x_2}{(x_1 + x_2)^3}\left[2 + 2(x_1 + x_2) + (x_1 + x_2)^2\right]e^{-(x_1 + x_2)} \quad (8.7)$$

and

$$f_3(x_1, x_2) = -Ei(-x_{\max}), \tag{8.8}$$

where $x_{\max} = $ maximum (x_1, x_2) and $Ei(x)$ is the exponential integral
function.

For normal marginals, the expressions become

$$f_2(x_1, x_2) = \frac{3}{\sqrt{\pi}\, 2^{3/2}} \frac{(x_1 x_2)^2}{(x_1^2 + x_2^2)^{5/2}}\left[1 - H\left(\frac{2}{x_1^2 + x_2^2}\right)\right] \tag{8.9}$$

where H is the gamma distribution function with shape parameter 5/2 and
scale parameter 1, and

$$f_3(x_1, x_2) = \begin{cases} 1 - \Phi(x_{\max}), & x_1 > 0, x_2 > 0 \\ \Phi(x_{\min}), & x_1 < 0, x_2 < 0 \\ 0, & x_1 x_2 < 0 \end{cases} \tag{8.10}$$

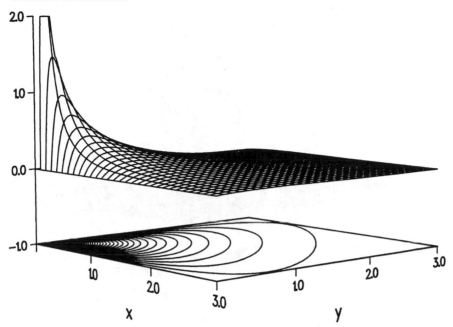

Figure 8.2. Identical uniforms, independent generators, with exponential marginals.

where Φ is the standard normal distribution function and $x_{\min}$ is the minimum of x_1 and x_2.

The individual cases listed above offer no mechanism for controlling the extent of dependence. To alleviate this deficiency, consider the density obtained by a probabilistic mixture of f_1, f_2, and f_3:

$$f(x_1, x_2) = qf_2(x_1, x_2) + pf_3(x_1, x_2) + (1 - p - q)f_1(x_1, x_2), \quad (8.11)$$

where $p + q \leqslant 1$; $p, q \geqslant 0$. Using the exponential marginal densities (8.7) and (8.8) in this form gives a distribution having correlation $p/2 + q/3$. For the normal marginals case (using (8.9) and (8.10) in (8.11)), the correlation is $3p/4$. Here the mixing parameter q does not enter the correlation explicitly, since (X_1, X_2) distributed according to f_2 has zero correlation (although it does not correspond to independence).

To assess the appearance of these densities, some contour and density plots are presented. Figure 8.2 illustrates density (8.7) having exponential marginals. Figures 8.3 and 8.4 conform to standard normal marginals.

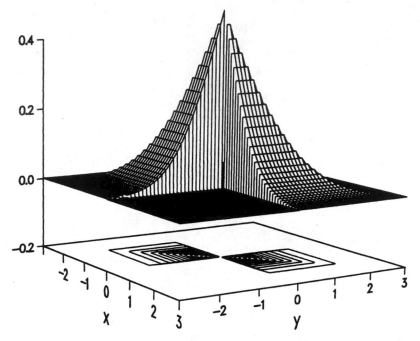

Figure 8.3. Independent uniforms, identical generators, with normal marginals.

The normal marginals cases illustrate discontinuities along the axes. To obtain "smoother" densities, the following model is appropriate:

$$X_1 = Z(2U_1 - 1)$$

$$X_2 = Z(2U_2 - 1), \qquad (8.12)$$

where Z is $\sqrt{\chi^2_{(3)}}$, the U_i's are uniform 0–1, and Z and (U_1, U_2) are independent. This model has the effect of spreading the mass of (U_1, U_2) on the unit square neatly through the larger square $[-1, 1] \times [-1, 1]$ and thus avoiding discontinuities on the axes. The variables X_1 and X_2 have marginal standard normal distributions. The joint density of X_1 and X_2 in (8.12) is

$$f(x_1, x_2) = \int_{\max |x_1|, |x_2|}^{\infty} (8\pi)^{-1/2} e^{-t/2} g\left[\frac{x_1}{2t} + \frac{1}{2}, \frac{x_2}{2t} + \frac{1}{2} \right] dt, \quad (8.13)$$

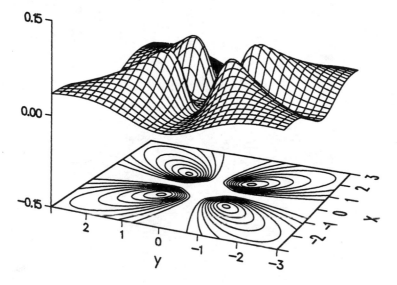

Figure 8.4. Identical uniforms, independent generators, with normal marginals.

where g is a bivariate uniform density governing (U_1, U_2). Two special cases are of interest. If g is constant (independent uniforms), then the density in (8.13) simplifies to

$$f(x_1, x_2) = \tfrac{1}{2}[1 - \Phi(\max |x_1|, |x_2|)]. \tag{8.14}$$

This is similar to the form in (8.10) but has mass in all quadrants and avoids the discontinuous density along the axes. Also, the components of (X_1, X_2) are uncorrelated but they are not independent.

More generally, if h is assumed to follow the Morgenstern uniform distribution (Section 10.1), then the density in (8.13) becomes

$$f(x_1, x_2) = \frac{\alpha x_1 x_2}{2A} \phi(A) + \frac{1 - \alpha x_1 x_2}{2}[1 - \Phi(A)], \tag{8.15}$$

where $A = \max |x_1|, |x_2|$ and α is the standard normal density function. Some views of this density for $\alpha = -1, 0$, and 1 are presented in Figures 8.5–8.7. The correlation corresponding to (8.15) is $\alpha/3$. Hence, only weak dependence is obtainable with this distribution. In spite of this small range of dependence, the unusual contours suggest that the distribution is appropriate as an alternative to the usual bivariate normal in robustness applications.

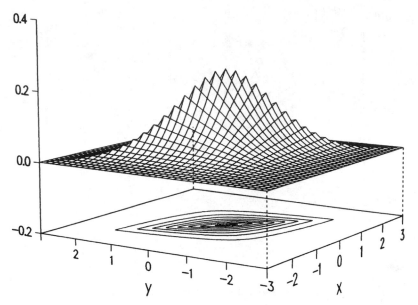

Figure 8.5. Хинчин normal, $\alpha = -1$.

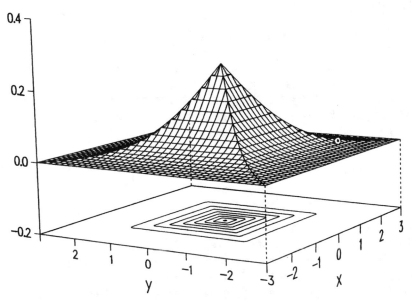

Figure 8.6. Хинчин normal, $\alpha = 0$.

158

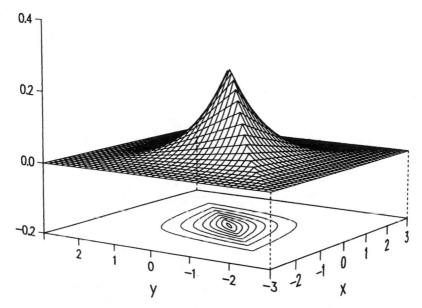

Figure 8.7. Хинчин normal, $\alpha = 1$.

A variety of distributions have been presented in this chapter. In all cases, variate generation is immediate from the construction schemes.

Multivariate ($p > 2$) forms have received only casual mention aside from (8.2). Variate generation does not pose a problem since the construction scheme can be directly implemented given a specified distribution. For the general case, the main limitation in (8.2) deals with the lack of suitable multivariate distributions having uniform 0–1 marginals. Of course, the multivariate normal distribution can be transformed to uniform marginals using the probability integral transform Φ. Other possibilities include the Burr, Pareto, logistic standard forms (Chapter 9) or the multivariate Morgenstern distribution (Section 10.1). Any multivariate distribution with uniform marginals can feed directly into the basic construction of this chapter.

Multivariate Burr, Pareto, and Logistic Distributions

The distributions presented in this chapter are not so closely attached to the multivariate normal distribution as the Johnson system, contaminated normal, or elliptically contoured distributions. A basic distributional form is given from which the multivariate Burr, Pareto, and logistic distributions are readily obtained. These distributions have historically been treated as important but seemingly isolated distributions. Section 9.1 describes the unified distribution having uniform marginal distributions and gives its basic properties. This chapter expands on the work of Cook and Johnson (1981, 1986), who also demonstrated that the form of the distribution with normal marginals provided excellent fits to data sets in a geological investigation. Additional generalizations of the unified distribution are pursued in Section 9.2. A form that incorporates Morgenstern's distribution seems particularly promising for Monte Carlo applications.

9.1. STANDARD FORM AND PROPERTIES

The multivariate Burr, Pareto, and logistic distributions have density functions and marginal distribution functions as indicated below:

Burr:

$$f(x_1, x_2, \ldots, x_n) = \frac{\Gamma(k+n)}{\Gamma(k)} \left[1 + \sum_{j=1}^{n} d_j x_j^{c_j} \right]^{-(k+n)} \prod_{j=1}^{n} \left(d_j c_j x_j^{c_j-1} \right),$$

$$x_j > 0 \quad (9.1)$$

$$F_{X_i}(x_i) = 1 - \left(1 + d_i x_i^{c_i} \right)^{-k}. \quad (9.2)$$

Pareto:

$$f(x_1, x_2, \ldots, x_n) = \frac{\Gamma(a+n)}{\Gamma(a)} \left(\prod_{j=1}^{n} \theta_j \right)^{-1} \left(\sum_{j=1}^{n} \theta_j^{-1} x_j - n + 1 \right)^{-(a+n)},$$

$$x_j > \theta_j > 0 \quad (9.3)$$

$$F_{X_i}(x_i) = 1 - \left(\frac{\theta_j}{x_i} \right)^a. \tag{9.4}$$

Logistic:

$$f(x_1, x_2, \ldots, x_n) = \frac{\Gamma(\alpha+n)}{\Gamma(\alpha)} \prod_{i=1}^{n} \exp(-x_i) \left[\sum_{i=1}^{n} \exp(-x_i) + 1 \right]^{-(\alpha+n)},$$

$$-\infty < x_i < \infty \quad (9.5)$$

$$F_{X_i}(x_i) = [1 + \exp(-x_i)]^{-\alpha}. \tag{9.6}$$

Interest in these distributions has grown in recent years. Takahasi (1965) developed the form in (9.1) and named it in honor of its univariate distribution originator (Burr, 1942). This bivariate Burr distribution was slightly generalized by Durling (1975). The multivariate Pareto has been described by Johnson and Kotz (1972) and applied in a medical study by Hutchinson (1979). The functional form in (9.5) gives Satterthwaite and Hutchinson's (1978) bivariate logistic form, which includes Gumbel's (1961) original simpler distribution. Satterthwaite and Hutchinson use the distribution to study four correlation measures in a disease transmission study.

The most obvious similarity among the three density functions is the proportionality constant $\Gamma(b+n)/\Gamma(b)$ where b is k, a, or α. On closer examination, notice that for $c_j = 1$ in the Burr distribution, its difference from the Pareto distribution consists of a location shift via the θ_j's. Further, for the logistic distribution, the transformation $y_i = \exp(-x_i)$ produces a form comparable to the Burr and Pareto distributions.

The key idea in recognizing the similar structure underlying the three distributions is to remove the effect of the different marginal distributions. This is accomplished by making the appropriate probability integral transformation to each component of the distributions:

$$U_i = F_i(X_i), \quad i = 1, 2, \ldots, n,$$

where F_i is the distribution function of the ith component of $\mathbf{X}$. This leads to multivariate distributions having uniform marginal distributions and density functions:

$$g_B(u_1, u_2, \ldots, u_n)$$

$$= \frac{\Gamma(k+n)}{\Gamma(k)k^n} \prod_{i=1}^{n} (1 - \mu_i)^{-(1/k)-1} \left[\sum_{i=1}^{n} (1 - \mu_i)^{-(1/k)} - n + 1 \right]^{-(k+n)}$$

(9.7)

$$g_p(u_1, u_2, \ldots, u_n)$$

$$= \frac{\Gamma(a+n)}{\Gamma(a)a^n} \prod_{i=1}^{n} (1 - \mu_i)^{-(1/a)-1} \left[\sum_{i=1}^{n} (1 - \mu_i)^{-1/a} - n + 1 \right]^{-(a+n)}$$

(9.8)

$$g_L(u_1, u_2, \ldots, u_n)$$

$$= \frac{\Gamma(\alpha+n)}{\Gamma(\alpha)a^n} \prod_{i=1}^{n} (u_i)^{-(1/\alpha)-1} \left[\sum_{i=1}^{n} u_i^{-(1/\alpha)} - n + 1 \right]^{-(\alpha+n)} \qquad (9.9)$$

The Burr density g_B and the Pareto density g_P are identical with k and a associated. The logistic density g_L is almost the same as the other two although u_i is in place of $1 - \mu_i$. Applying the transformation $V_i = 1 - U_i$, $i = 1, 2, \ldots, n$, in (9.9) preserves the uniform marginal distributions and yields a form identical to (9.7) and (9.8) with α associated with k and a.

The standard form of the unified multivariate Burr-Pareto-logistic distribution is defined to have density function

$$f_n(u_1, u_2, \ldots, u_n) = \frac{\Gamma(\alpha+n)}{\Gamma(\alpha)\alpha^n} \prod_{i=1}^{n} u_i^{-(1/\alpha)-1} \left[\sum_{i=1}^{n} u_i^{-1/\alpha} - n + 1 \right]^{-\alpha-n}$$

(9.10)

for $0 \leqslant u_i \leqslant 1$, $i = 1, 2, \ldots, n$, and zero elsewhere. Each of the forms (9.1), (9.3), and (9.5) are directly obtained from (9.10) via simple componentwise transformations. The corresponding distribution function of (9.10) is

$$F_n(u_1, u_2, \ldots, u_n) = \left[\sum_{i=1}^{n} u_i^{-1/\alpha} - n + 1 \right]^{-\alpha}, \qquad (9.11)$$

where $0 \leqslant u_i \leqslant 1$, $i = 1, 2, \ldots, n$. Although it is evident from the construction of (9.10) via the probability integral transform that the marginal distributions are uniform 0–1, a simple verification of this result can also be made by evaluating F_n at $u_2 = u_3 = \cdots = u_n = 1$ in (9.11). A result of theoretical interest is the conditional expectation of U_1 given $U_2 = u_2, \ldots, U_n = u_n$:

$$E(U_1 | U_2 = u_2, \ldots, U_n = u_n)$$

$$= \frac{\alpha + n - 1}{(n + 2 - 1)K} H\left(\alpha + n, 1; n + 2\alpha; \frac{K - 1}{K}\right),$$

where $K = \sum_{i=2}^{n} u_i^{-1/\alpha} - (n - 2)$ and H is the hypergeometric function (Johnson and Kotz, 1969, p. 8) defined by

$$H(a; b; c; d) = \sum_{i=0}^{\infty} \frac{a^{[i]} b^{[i]} d^{[i]}}{c^{[i]} i!}.$$

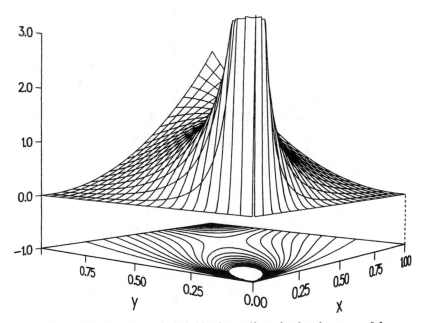

Figure 9.1. Burr-Pareto-logistic bivariate uniform density plot, $\alpha = +0.5$.

The notation $a^{[i]}$ refers to the ascending factorial defined by $a^{[i]} = a(a + 1) \cdots (a + i - 1)$.

The parameter α is positive and governs the strength of dependence for any pair U_i and U_j ($i \neq j$). The dependence is the same for all pairs of variables, a limiting characteristic that Section 9.2 attempts to overcome. To interpret α, it can be easily verified that in the bivariate case, for example,

$$\lim_{\alpha \to \infty} F_2(U_1, U_2; \alpha) = U_1 U_2 \tag{9.12}$$

$$\lim_{\alpha \to 0} F_2(U_1, U_2; \alpha) = \min(U_1, U_2), \tag{9.13}$$

which correspond to independence (9.12) and complete dependence (9.13). In fact, the distribution in (9.13) assigns all mass to the line segment $U_1 = U_2$ in the unit square. The correlation coefficient ρ of U_1 and U_2 is

$$\rho = 12 E(U_1 U_2) - 3, \tag{9.14}$$

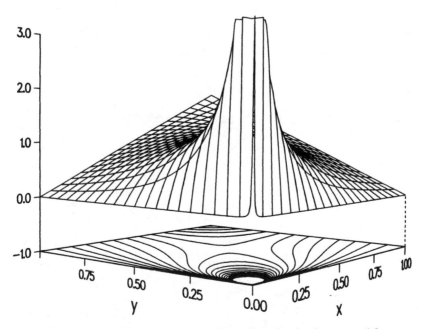

Figure 9.2. Burr-Pareto-logistic bivariate uniform density plot, $\alpha = +1.0$.

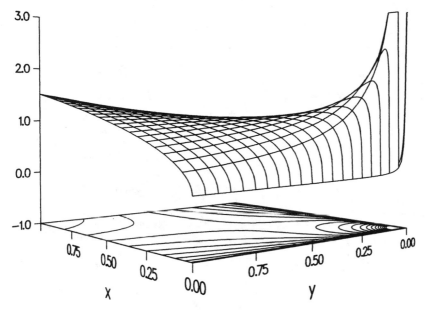

Figure 9.3. Burr-Pareto-logistic bivariate uniform density plot, $\alpha = +2.0$.

where

$$E(U_1 U_2) = \frac{\Gamma^2(2\alpha + 1)}{\Gamma(\alpha)} \sum_{j=0}^{n} \frac{\Gamma(a + j + 2)j!}{\Gamma^2(2\alpha + j + 2)}.$$

From (9.12) and (9.13) the limiting cases are $\rho \to 0$ as $\alpha \to \infty$ and $\rho \to 1$ as $\alpha \to 0$. In the standard form (9.10), all correlations are positive and equal. To obtain negative correlations, some but not all of the U_i's are transformed to $1 - U_i$. In the bivariate case, the distribution $(U_1, 1 - U_2)$ represents a $90°$ rotation of the distribution (U_1, U_2), so that $\rho(U_1, 1 - U_2) = -\rho(U_1, U_2)$. It should be emphasized that (U_1, U_2) and $(1 - U_1, 1 - U_2)$ do not have the same distributions, since the joint distributions are not symmetric about the line $U_2 = 1 - U_1$.

Some of the results stated above can be reinforced by the bivariate density and contour plots in Figures 9.1–9.5. Figures 9.1–9.3 correspond to the bivariate uniform density

$$f(u_1, u_2) = \frac{\alpha + 1}{\alpha} (u_1 u_2)^{-(1/\alpha)-1} \left(u_1^{-1/\alpha} + u_2^{-1/\alpha} - 1 \right)^{-(\alpha+2)} \quad (9.15)$$

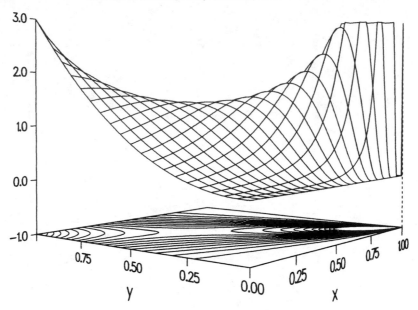

Figure 9.4. Burr-Pareto-logistic bivariate uniform density plot with $U_1 \rightarrow (1 - U_1)$ transformation, $\alpha = +0.5$.

with α selected as indicated. Since f tends to infinity as $u_1 u_2$ tend to zero, for plotting purposes values of f greater than 3 were set to 3 (thus explaining the mesa near the origin). The density is bimodal with a saddle point. A different viewing angle is selected in Fig. 9.3. Figures 9.4 and 9.5 represent the transformations $(U_1, 1 - U_2)$ and $(1 - U_1, 1 - U_2)$ of the distribution (U_1, U_2) with $\alpha = 0.5$, as in Figure 9.3. These views are from the same observation point as in Figure 9.1.

Comparisons with the distributions of previous chapters are facilitated by considering the standard form (9.15) transformed to normal marginal distributions, having the density function

$$g(x_1, x_2) = \frac{\alpha + 1}{\alpha} \phi(x_1)\phi(x_2)[\Phi(x_1)\Phi(x_2)]^{-(1/\alpha)-1}$$

$$\times \left\{ [\Phi(x_1)]^{-1/\alpha} + [\Phi(x_2)]^{-1/\alpha} - 1 \right\}^{-(\alpha+2)} \quad (9.16)$$

where ϕ is the standard normal density function and Φ is the standard normal distribution function. Figures 9.6–9.9 offer density and contour

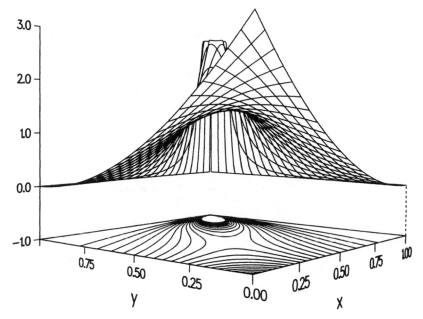

Figure 9.5. Burr-Pareto-logistic bivariate uniform density plot with $U_1 \to (1 - U_1)$ and $U_2 \to (1 - U_2)$ transformations, $\alpha = +0.5$.

plots for four choices of α as indicated. Unlike the usual bivariate normal distribution, the contours within a figure are not similar geometrical structures. Contours near the mode are elliptical or for low correlations nearly spherical, whereas those farther out are pointed or blunt at opposite ends of the main diagonal. Of course, near the mode, the density should have elliptical contours, as the function is "smooth" and can be approximated by a quadratic function. It is also apparent from the figures that the variance of the conditional distributions X_2 given $X_1 = x_1$ increases with x_1.

To apply this distribution in Monte Carlo work, a variate generation algorithm is required. The standard form (9.10) can be obtained directly from independent variates. Let $Y_1, Y_2, \ldots, Y_n$ be independent and identically distributed exponential variates and let X be an independent $\Gamma(\alpha, 1)$. The joint density function of

$$U_i = \left[1 + \frac{Y_i}{X} \right]^{-\alpha}$$

is given by (9.10). Gamma and exponential variate generation is easy given

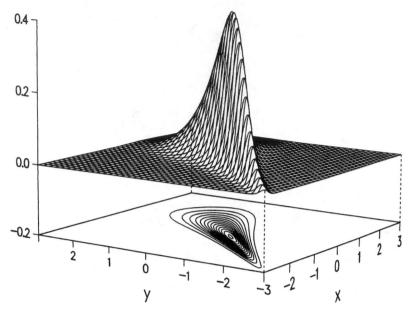

Figure 9.6. Burr-Pareto-logistic normal density and contour plots, $\alpha = 0.25$.

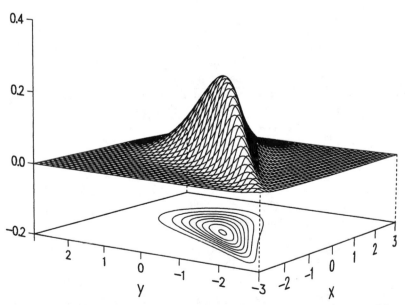

Figure 9.7. Burr-Pareto-logistic normal density and contour plots, $\alpha = 0.5$.

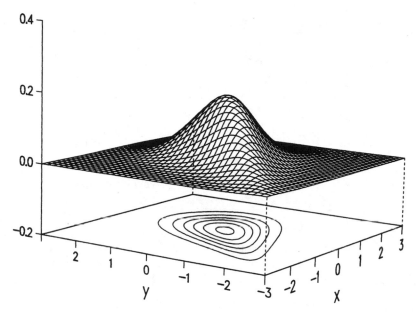

Figure 9.8. Burr-Pareto-logistic normal density and contour plots, $\alpha = 1$.

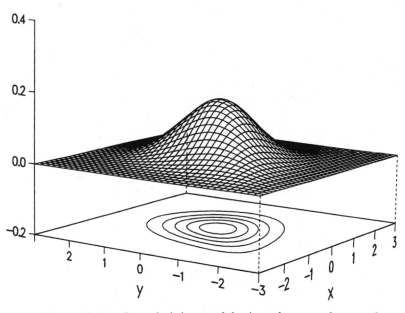

Figure 9.9. Burr-Pareto-logistic normal density and contour plots, $\alpha = 2$.

169

TABLE 9.1. Variate Generation Formulas

Marginal Distributions	Standard Form	Translated Standard Form
Uniform	$(1 + Y_i/X)^{-\alpha}$ (9.10)	$1 - (1 + Y_i/X)^{-\alpha}$
Logistic	$-\ln(Y_i/X)$ (9.5)	$-\ln\{[1 - (1 + Y_i/X)^{-\alpha}]^{-1/\alpha} - 1\}$
Burr	$[d_i^{-1}\{[1 - (1 + Y_i/X)^{-\alpha}]^{-1/\alpha} - 1\}]^{1/c_i}$	$(d_i^{-1}Y_i/X)^{1/c_i}$ (9.1)
Pareto	$\theta_i[1 - (1 + Y_i/X)^{-\alpha}]^{-1/\alpha}$	$\theta_i(1 + Y_i/X)$ (9.3)
Normal	$\Phi^{-1}[(1 + Y_i/X)^{-\alpha}]$	$\Phi^{-1}[1 - (1 + Y_i/X)^{-\alpha}]$

Table 1 from Cook, R. D. and Johnson, M. E. (1981). "A Family of Distributions for Modelling Non-Elliptically Symmetric Multivariate Data." *Journal of the Royal Statistical Society, Series B*, **43**, 210–218.

the methods in Section 2.5. Formulas for some special cases considered in this chapter are summarized in Table 9.1. The translated standard form involves replacing U_i by $1 - U_i$, $i = 1, 2, \ldots, n$, in (9.10).

9.2. GENERALIZATIONS

The distribution presented in the previous section would seen to be of primary interest in two or possibly three dimensions in view of the equicorrelation structure. This means that the single dependence parameter α implies that $\rho(U_i, U_j) = \pm\rho(U_k, U_l)$ for all $i \neq j$, $k \neq l$. The correlations can differ in sign but not absolute value if some of the U_i's are replaced by $1 - U_i$ in the standard form. In this section, some attempts are made to broaden the allowable dependence structure in the distribution. An obvious strategy is to use dependent exponential variates $Y_1, Y_2, \ldots, Y_n$ when forming $(1 + Y_i/X)^{-\alpha}$, where X is an independent $\Gamma(\alpha, 1)$. Since many nontrivial multivariate exponential distributions are available, this tactic would appear fruitful. The simple case with (Y_1, Y_2) having Morgenstern's (Section 10.1) distribution with exponential marginals is pursued in detail. Some other possibilities for multivariate exponential distributions are then briefly mentioned.

One possible generalization to the standard form (9.10) employs Morgenstern's exponential distribution for (Y_1, Y_2) having density function

$$f(y_1, y_2) = e^{-y_1 - y_2}[1 + \beta(2e^{-y_1} - 1)(2e^{-y_2} - 1)],$$

$$y_i > 0, \quad -1 \leqslant \beta \leqslant 1. \quad (9.17)$$

This distribution is Morgenstern's bivariate distribution (Section 10.1) with exponential marginal distributions. Let (Y_1, Y_2) have density f, and let X be an independent $\Gamma(\alpha, 1)$ variate. The pair $Z_1 = Y_1/X$, $Z_2 = Y_2/X$ has a bivariate distribution with Burr marginal distributions corresponding to density function

$$g(z_1, z_2) = \alpha(\alpha + 1)\big[(\beta + 1)(z_1 + z_2 + 1)^{-\alpha-2}$$

$$+ 4\beta(2z_1 + 2z_2 + 1)^{-\alpha-2} - 2\beta(2z_1 + z_2 + 1)^{-\alpha-2}$$

$$- 2\beta(z_1 + 2z_2 + 1)^{-\alpha-2}\big], \qquad z_1, z_2 \geqslant 0.$$

The transformation $U_1 = (1 + Z_1)^{-\alpha}$, $U_2 = (1 + Z_2)^{-\alpha}$ converts to a bivariate distribution with uniform 0–1 marginals and density function:

$$h(u_1, u_2) = \frac{\alpha + 1}{\alpha}(u_1 u_2)^{-(1/\alpha)-1}$$

$$\times \Big\{(\beta + 1)\big[u_1^{-(1/\alpha)} + u_2^{-(1/\alpha)} - 1\big]^{-\alpha-2}$$

$$+ 4\beta\big[2u_1^{-(1/\alpha)} + 2u_2^{-(1/\alpha)} - 3\big]^{-\alpha-2}$$

$$- 2\beta\big[2u_1^{-(1/\alpha)} + u_2^{-(1/\alpha)} - 2\big]^{-\alpha-2}$$

$$- 2\beta\big[u_1^{-(1/\alpha)} + 2u_2^{-(1/\alpha)} - 2\big]^{-\alpha-2}\Big\} \qquad (9.18)$$

and corresponding distribution function

$$H(u_1, u_2) = (\beta + 1)\big[u_1^{-(1/\alpha)} + u_2^{-(1/\alpha)} - 1\big]^{-\alpha}$$

$$+ \beta\big[2u_1^{-(1/\alpha)} + 2u_2^{-(1/\alpha)} - 2\big]^{-\alpha}$$

$$- \beta\big[2u_1^{-(1/\alpha)} + u_2^{-(1/\alpha)} - 2\big]^{-\alpha}$$

$$- \beta\big[u_1^{-(1/\alpha)} + 2u_2^{-(1/\alpha)} - 2\big]^{-\alpha}. \qquad (9.19)$$

Some special cases of this distribution warrant comment. For $\beta = 0$, independent exponentials, the bivariate standard form of Section 9.1 is obtained. The limiting case in (9.18) as $\alpha \to 0$ and for any $\beta \in [-1, 1]$ is the degenerate distribution corresponding to $U_1 = U_2$. The limiting density function of (9.18) as $\alpha \to \infty$ is

$$h(u_1, u_2) = 1 + \beta(2u_1 - 1)(2u_2 - 1), \qquad (9.20)$$

which is Morgenstern's uniform distribution.

Further revelations of the role of α and β in (9.18) are provided in Figure 9.10. The middle row of plots corresponds to the distribution (9.15) of the previous section. The effects of β are most noticeable for large values of α, moving the saddle point markedly. It is somewhat easier to assess the role of β using the density function with standard normal marginals

$$p(x_1, x_2) = \phi(x_1)\phi(x_2)h\big[\Phi(x_1), \Phi(x_2)\big], \qquad (9.21)$$

where h is from (9.18), ϕ is the standard normal density, and Φ is the standard normal distribution function. Figure 9.11 provides multiple contour plots for this normal marginals case. Nonelliptical contours arise in all cases away from the mode. The distributions are clearly symmetric about the line $x_1 = x_2$ but very asymmetric about $x_1 = -x_2$. Increasing conditional variances of X_2 given $X_1 = x_1$ are apparent in the figure. As a final look at the distribution, Figure 9.12 provides a curious set of plots for (9.15) transformed to have standard exponential marginals. Independence would correspond to contours that are parallel lines with slope equal to -1. The $\alpha = 10$, $\beta = 0$ case exhibits this situation somewhat.

In some Monte Carlo applications, such as the discriminant analysis studies mentioned in Chapter 5, it is helpful to be able to specify the correlation structure in a population. Table 9.2 provides estimated values of the correlations over a broad grid of α-β combinations for normal marginals. These values were based on one million generated pairs of each α-β case, leading to a standard error of at most 0.001 ($\sqrt{(1 - \rho^2)/n}$). For the extreme case $\alpha = \infty$, the correlation is β/π for normal marginals and $\beta/3$ for uniform marginals. The correlation $\rho(X_1, X_2)$ is but one measure of the degree of association between random variables. It captures the extent to which X_1 and X_2 are linearly related. The cases with $\rho(X_1, X_2)$ equal to -1 and 1 can be interpreted most easily. Figure 9.13 gives a plot of the case (9.21) with $\alpha = 3$, $\beta = -1$. The correlation is -0.009, or approximately zero. The distribution does not correspond to independence, however, since the contours are not concentric circles.

Variate generation for the distribution in (9.18) is straightforward from the construction underlying the distribution. First, a pair (V_1, V_2) having Morgenstern's bivariate uniform distribution is generated, as described in Section 10.1. Independently, a gamma $\Gamma(\alpha, 1)$ variate X is generated (Section 2.5). The desired pair (U_1, U_2) distributed as (9.18) is obtained by

$$U_i = \big[1 - X^{-1}\ln(1 - V_i)\big]^{-\alpha}, \qquad i = 1, 2.$$

Specified marginal distributions with distribution functions F_1 and F_2 can be obtained in the usual way via $[F_1^{-1}(U_1), F_2^{-1}(U_2)]$.

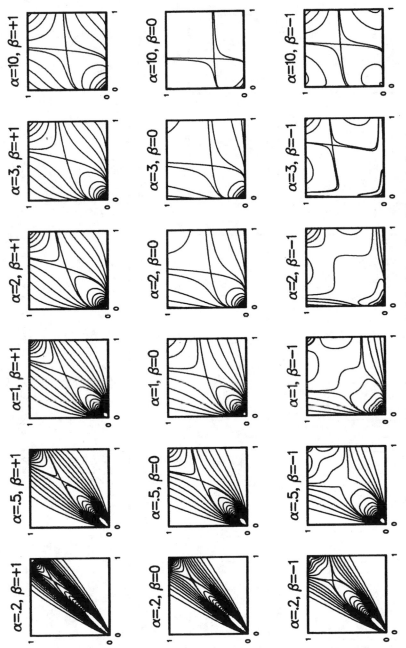

Figure 9.10. Generalized Burr-Pareto-logistic plots, uniform 0–1 marginals.

173

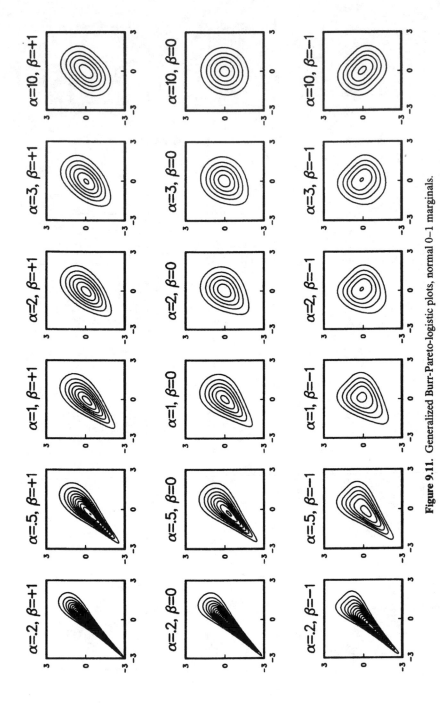

Figure 9.11. Generalized Burr-Pareto-logistic plots, normal 0–1 marginals.

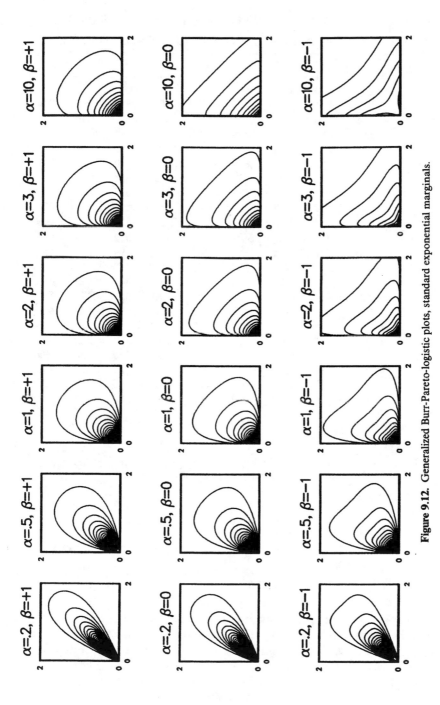

Figure 9.12. Generalized Burr-Pareto-logistic plots, standard exponential marginals.

175

TABLE 9.2. Correlation as a Function of α and β, Normal Marginals

α β:	−1	−.75	−.5	−.25	0	.25	.5	.75	1.
0	1.000	1.000	1.000	1.000	1.000	1.000	1.000	1.000	1.000
0.2	.822	.835	.843	.853	.865	.869	.882	.890	.901
0.3	.736	.750	.766	.781	.797	.810	.826	.839	.855
0.4	.662	.681	.698	.718	.740	.752	.774	.792	.811
0.5	.590	.613	.638	.663	.687	.707	.731	.754	.777
0.6	.533	.558	.583	.609	.640	.663	.690	.717	.747
0.7	.475	.507	.538	.571	.597	.629	.658	.689	.716
0.8	.431	.462	.496	.527	.562	.592	.625	.657	.690
0.9	.385	.419	.457	.491	.530	.563	.597	.631	.670
1.0	.348	.384	.420	.459	.498	.537	.576	.612	.649
1.2	.278	.321	.360	.407	.450	.491	.533	.573	.616
1.4	.225	.270	.314	.362	.407	.451	.499	.544	.587
1.6	.176	.224	.274	.323	.372	.421	.469	.519	.568
1.8	.138	.187	.239	.290	.342	.394	.445	.496	.548
2.0	.102	.157	.209	.263	.316	.370	.425	.478	.534
2.5	.034	.093	.151	.207	.268	.325	.382	.439	.499
3	−.009	.048	.071	.170	.229	.291	.350	.415	.474
4	−.082	−.015	.049	.115	.178	.247	.310	.379	.442
10	−.216	−.145	−.071	.004	.077	.151	.226	.298	.372
20	−.266	−.190	−.112	−.038	.041	.116	.193	.269	.348
40	−.294	−.217	−.136	−.058	.020	.098	.177	.253	.333
∞	−.318	−.239	−.159	−.080	.000	.080	.159	.239	.318

Multivariate extensions of the form in (9.18) can proceed using the multivariate Morgenstern distribution described in Section 10.1. An explicit construction is:

1. Generate **U** having density (10.3).
2. Set $Y_i = -\ln(1 - U_i)$, $i = 1, 2, \ldots, n$, to obtain exponential variates.
3. Set $Z \sim \Gamma(\alpha, 1)$, independently of **X**.
4. Set $V_i = 1 - (1 + Y_i/Z)^{-\alpha}$, $i = 1, 2, \ldots, n$.

The density for **V** can be worked out explicitly although the bookkeeping becomes onerous as n gets large. The range of correlations possible for each (V_i, V_j) is restricted by the choice of α. For example, for $\alpha = 1$ and normal marginal distributions, $0.348 \leq \rho(X_i, X_j) \leq 0.649$ for any choice of $i \neq j$. Although this is an improvement over the distribution in Section 9.1, a wider range could be needed in some applications.

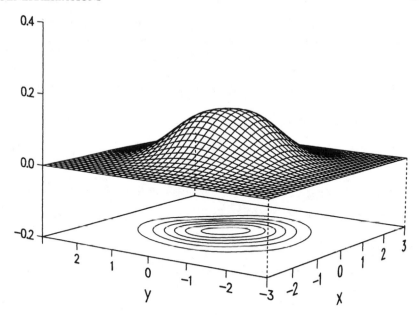

Figure 9.13. Generalized Burr-Pareto-logistic plots, normal maginals, $\alpha = 3$, $\beta = -1$.

The use of other multivariate exponential distributions in the construction phase could circumvent this problem at the expense of dealing with more complicated functional forms. Some possible multivariate exponential distributions for this purpose are now mentioned.

One easy construction scheme (Johnson and Kotz, 1972, p. 217) uses

$$Y_i = X_0 + X_i$$

where the X_i's $(i = 0, 1, \ldots, p)$ are independent gamma variates $\Gamma(\alpha_i, 1)$ with $\alpha_0 + \alpha_i = 1$. The density function of $(Y_1, \ldots, Y_p)$ requires numerical integration to be evaluated.

Another construction scheme has been examined by Jensen (1970). Let $\mathbf{X}$ be $N_p(\mathbf{0}, \Sigma_1)$ and independent of $\mathbf{Z}$, which is $N_p(\mathbf{0}, \Sigma_2)$. Form $\mathbf{Y}$ having ith component $X_i^2 + Z_i^2$. Of course, the Y_i's are exponential (provided the variance of X_i equals the variance of Z_i), but as in the preceding construction, the density function of $\mathbf{Y}$ is complex.

Gumbel (1960) has examined a variety of bivariate exponential distributions, one of which is examined in more detail in Section 10.3. Another possibility is given by the distribution function

$$F(x, y) = 1 - \exp(-x) - \exp(-y) + \exp\left[-(x^\alpha + y^\alpha)^{1/\alpha}\right].$$

Freund's (1961) bivariate exponential distribution ought to be mentioned, since with suitable restrictions on the parameters of the distributions, both marginal distributions are exponential.

Perhaps the most promising candidate multivariate exponential distribution is the one developed by Raftery (1984, 1985). In the bivariate case, the model is

$$X_1 = (1 - \pi_1)Y_1 + I_1 Z$$
$$X_2 = (1 - \pi_2)Y_2 + I_2 Z \tag{9.21}$$

where Y_1, Y_2, and Z are independent exponential random variables, I_1 and I_2 are discrete random variables with joint distribution

$$P[I_1 = j, I_2 = k] = P_{jk} \text{ for } j, k = 0, 1,$$

$\pi_i = P[I_i = 1]$, $i = 1, 2$.

This setup leads to the variate (X_1, X_2), which has exponential components that are positively correlated (the correlation is $2p_{11} - \pi_1 \pi_2$). Negative correlations are obtained via the easy adjustment to (9.21):

$$X_1 = (1 - \pi_1)Y_1 - I_1 \ln U$$
$$X_2 = (1 - \pi_2)Y_2 - I_2 \ln(1 - U), \tag{9.22}$$

where U is uniform 0–1 and independent of Y_1 and Y_2. For (9.22), the correlation is $\rho(X_1, X_2) = (\pi^2/6)p_1 - \pi_1 \pi_2$. By the above construction scheme, variate generation is immediate. Moreover, Raftery provides both a reliability model and shock model for which (9.21) holds.

A convenient multivariate extension of (9.21) is

$$X_i = (1 - \pi_i)Y_i + Z_{J_i}, \qquad i = 1, \ldots, n, \tag{9.23}$$

where $Y_1, \ldots, Y_n$; $Z_1, \ldots, Z_m$ are independent standard exponential variates,
$(J_1, \ldots, J_n)$ is a random vector with discrete components, each having support $\{0, 1, \ldots, m\}$,
the marginal distributions of the J_i's are given by

$$P[J_i = 0] = 1 - \pi_i$$
$$P[J_i = k] = \pi_{ik}, \qquad i = 1, \ldots, n; \, k = 1, \ldots, m$$

$$\pi_i = \sum_{k=1}^{m} \pi_{ik}.$$

Note that Z_0 is defined to be identically zero. The multivariate distribution given in (9.23) has exponential components, is easy to generate, and has readily specified correlation structure. Raftery has determined pragmatic, parsimonious forms of this distribution having $(n - 1)^2$ parameters in $n \geqslant 4$ dimensions.

Of the forms mentioned above, Raftery's distribution seems the most promising for possible use in constructing Burr type distributions as described in this section. Raftery's distribution is of interest in its own right among the multivariate distributions having exponential components.

CHAPTER 10

Miscellaneous Distributions

The distributions in this chapter are included as they have attracted some attention in multivariate simulation studies. While the distributions covered in earlier chapters are probably better suited to the investigations mentioned in Chapter 1, these miscellaneous distributions are nevertheless worth considering. The Morgenstern distribution (Section 10.1) has weak dependence properties but was found useful in expanding the structure of the Burr-Pareto-logistic class in Section 9.2. The Plackett distribution (Section 10.2) has some interesting features—especially with regard to its genesis. Also included in this chapter are the Gumbel (Section 10.3) and Ali-Mikhail-Haq (Section 10.4) distributions. Finally, the Wishart distribution is considered in Section 10.5. Although it is primarily of interest in normal theory multivariate analysis, the Wishart distribution has spawned some interesting variate generation work.

10.1. MORGENSTERN'S DISTRIBUTION

A distribution usually credited to Morgenstern (1956) has received remarkable attention, especially in view of its limited modeling capability. Another intriguing aspect of the distribution is the nonconstant set of contributors whose names are occasionally attached to it. Eyraud (1938) seems to have the earliest publication that presents the basic distributional form, given later by Morgenstern (1956). Farlie (1960) generalized the distribution slightly while Gumbel (1958) examined its special case with exponential marginals. Both Farlie and Gumbel cited the work of Morgenstern. For consistency with most references, the distribution to be given in this section is called Morgenstern's distribution, although Eyraud's earlier work should not be overlooked. Morgenstern's distribution has the bivariate density function

$$f(u_1, u_2) = 1 + \alpha(2u_1 - 1)(2u_2 - 1), \qquad 0 \leqslant u_1, u_2 \leqslant 1, -1 \leqslant \alpha \leqslant 1.$$

$$(10.1)$$

This distribution has uniform 0–1 marginals and the dependence structure between U_1 and U_2 is controlled by the parameter α. In particular the Pearson product moment correlation is $\alpha/3$ implying the limited range $-\frac{1}{3}$ to $\frac{1}{3}$.

A more general form with arbitrary marginal distribution functions G_1 and G_2 is given by (Johnson and Kotz, 1975)

$$G(x_1, x_2) = G_1(x_1)G_2(x_2)[1 + \alpha(1 - G_1(x_1))(1 - G_2(x_2))],$$

where G is the joint distribution function of (X_1, X_2). Assuming the corresponding densities g_1 and g_2 exist, the following form can be used:

$$g(x_1, x_2) = g_1(x_1)g_2(x_2)\{1 + \alpha[2G_1(x_1) - 1][2G_2(x_2) - 1]\}. \quad (10.2)$$

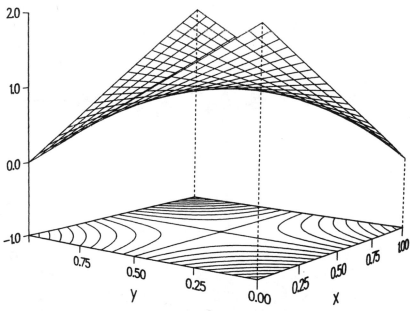

Figure 10.1. Morgenstern density, uniform marginals, $\alpha = +1$.

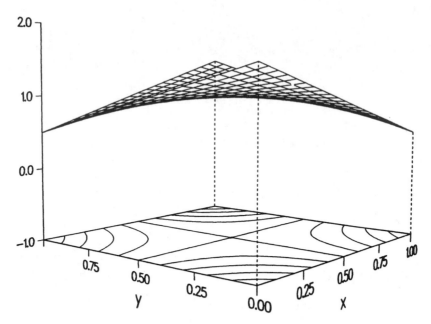

Figure 10.2. Morgenstern density, uniform marginals, $\alpha = +0.5$.

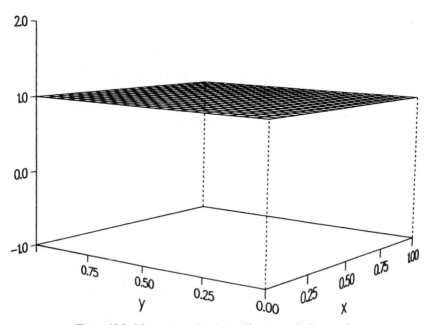

Figure 10.3. Morgenstern density, uniform marginals, $\alpha = 0$.

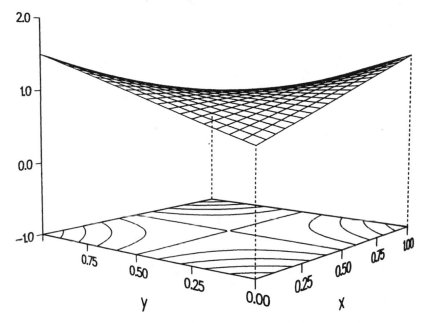

Figure 10.4. Morgenstern density, uniform marginals, $\alpha = -0.5$.

An understanding of the functional forms (10.1) and (10.2) can be gleaned from Figures 10.1–10.5. In all of these cases, the pictures give the density surface as viewed in three dimensions and the corresponding density contours at equally spaced increments of $g(x_1, x_2)$ projected onto a plane below the density. Thus, in Figure 10.1, the Morgenstern density having uniform marginals and $\alpha = +1$ is depicted. The nearest part of the surface is in fact the underside of the density function. The dashed bedpost lines are given to help establish the position of the contours of f relative to the surface itself. The density is identically 1 if $x = 0.5$ or $y = 0.5$. The point $(0.5, 0.5)$ is a saddle point of the Morgenstern uniform density. Figures 10.1–10.5 all correspond to uniform 0–1 marginal distributions. For $\alpha = 0$ (Figure 10.3), no contours are given since the density is flat, which corresponds to independent components. Viewed pairwise, the cases $\alpha = \pm 1$ (Figures 10.1 and 10.5) and $\alpha = \pm 0.5$ (Figures 10.2 and 10.4) represent the same surfaces if rotated 90 degrees $(0.5, 0.5)$.

Figures 10.6–10.8 represent standard normal marginal distributions used in the density g in (10.2). The case $\alpha = -1$ is clearly a rotation of the $\alpha = +1$ case. The situation with $\alpha = 0$ is the independent normal components distribution. The contours look roughly elliptical, which suggests a closeness to the usual bivariate normal.

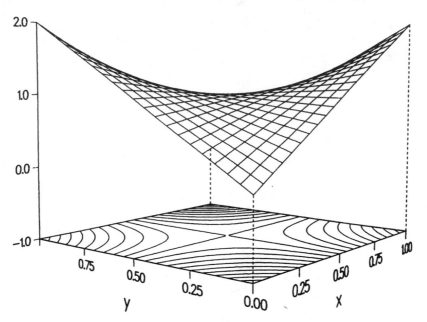

Figure 10.5. Morgenstern density, uniform marginals, $\alpha = -1$.

Figures 10.9–10.11 represent the exponential marginal distribution case of (10.2). Here, the positive and negative α values do not correspond to mere rotations in the distributions.

Random variates from the density in (10.2) are readily obtained by first generating (U_1, U_2) according to (10.1) and then applying $X_1 = G_1^{-1}(U_1)$ and $X_2 = G_2^{-1}(U_2)$. The conditional distribution approach as described in Section 3.1 is applicable for generating a pair (U_1, U_2) from (10.1). First, U_1 is generated as a uniform 0–1 variate. Next, the conditional distribution of U_2 given $U_1 = u_1$ has distribution function

$$F(u) = \int_0^u 1 + \alpha(2u_1 - 1)(2u_2 - 1)\, du_2$$

$$= \left[1 - \alpha(2u_1 - 1)\right]u + \alpha(2u_1 - 1)u^2,$$

which is quadratic in u. For $0 < p < 1$, the equation $F(U) = p$ has one root U in the unit interval. The appropriate calculations for generating

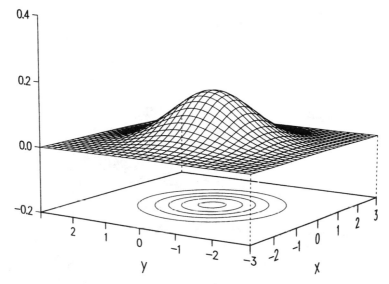

Figure 10.6. Morgenstern density, normal marginals, $\alpha = 0.0$.

from (10.1) are summarized as follows:

1. Generate V_1, V_2 independent uniform 0–1 and set $U_1 = V_1$.
2. Compute
   $$A = \alpha(2U_1 - 1) - 1$$
   $$B = [1 - 2\alpha(2U_1 - 1) + \alpha^2(2U_1 - 1)^2 + 4\alpha V_2(2U_1 - 1)]^{1/2}.$$
3. Set $U_2 = 2V_2/(B - A)$.

Although Morgenstern's distribution is easy to generate, its role in simulation work is generally limited to situations with weak dependence. The work of Schucany, Parr, and Boyer (1978) is useful in quantifying these limitations. Some of their key mathematical results and corresponding interpretations are now summarized.

Product-Moment Correlation

The product moment correlation coefficient ρ corresponding to the general form in (10.2) satisfies $|\rho| \leqslant \frac{1}{3}$ with equality occurring for uniform marginals. In other words, the maximum correlation over all choices of G_1 and G_2 is $\frac{1}{3}$ ($\alpha = 1$). For normal marginals, the maximum is $1/\pi$; for exponential marginals it is $\frac{1}{4}$; for double exponential marginals, the limit is 0.281.

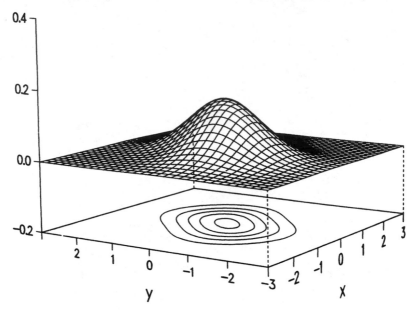

Figure 10.7. Morgenstern density, normal marginals, $\alpha = 1.0$.

Conditional Expectations

In general, if (X_1, X_2) is distributed according to (10.2),

$$E(X_1|X_2 = x_2) = E(X_1) + \alpha\{1 - 2G_2(x_2)\} \cdot \int x_1\{1 - 2G_1(x_1)\}\, dG_1(x_1),$$

which is linear in $G_2(x_2)$. Moreover,

$$\frac{\text{Var}\{E(X_1|X_2 = x_2)\}}{\text{Var}(X_1)} \leq \frac{\alpha^2}{3},$$

which indicates that at most one-third of the variance of X_1 can be attributed to X_2.

Probabilities of Rectangles

Let B denote a rectangular region in the plane and define $P_\alpha(B) = P[(X_1, X_2) \in B|\alpha]$. The following inequality is true for any (X_1, X_2) dis-

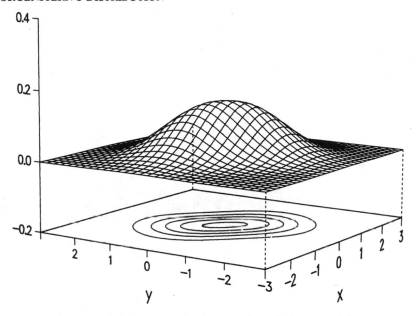

Figure 10.8. Morgenstern density, normal marginals, $\alpha = -1.0$.

tributed according to density (10.2):

$$|P_0(B) - P_\alpha(B)| \leqslant \frac{|\alpha|}{16}.$$

This inequality gives an indication of the closeness of probabilities using arbitrary α to probabilities based on independent ($\alpha = 0$) components. In contrast, for the bivariate normal distribution (Chapter 4) it is easy to construct a case where the right-hand side is $\frac{1}{4}$ using identical X_i's and the second quadrant

$$B = \{(x_1, x_2): x_1 < 0 \text{ and } x_2 > 0\}.$$

Although the work of Schucany and colleagues reveals explicitly some severe limitations of Morgenstern's distribution, this is not intended to suggest that the distribution should be scrapped. Morgenstern's distribution can certainly be used in Monte Carlo studies if weak dependence but not independence is a characteristic of interest. Also, if a new procedure breaks down with Morgenstern's distribution, then it will probably have diffi-

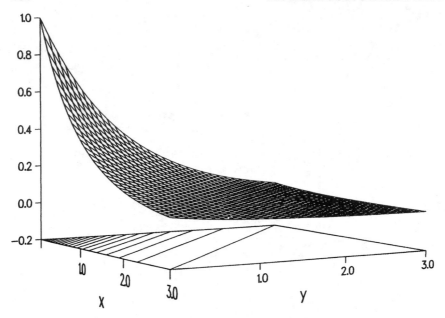

Figure 10.9. Morgenstern density, exponential marginals, $\alpha = 0$.

culties with other cases that admit stronger dependence. Another use of Morgenstern's distribution was given in Section 9.2, where it was used to generalize the Burr-Pareto-logistic class of distributions. Finally, Morgenstern's distribution is a first order approximation to Plackett's and Ali-Mikhail-Haq's distributions, which are described in Sections 10.2 and 10.4, respectively.

Huang and Kotz (1984) have indicated the following density form to expand the correlation structure of the Morgenstern distribution:

$$f(u, v) = 1 + \alpha(1 - 2u)(1 - 2v) + \beta uv(2 - 3u)(2 - 3v),$$

with $0 \leqslant u, v \leqslant 1$, $|\alpha| \leqslant 1$, $\alpha + \beta \leqslant 1$, $\beta \leqslant [3 - \alpha + (9 - 6\alpha - 3\alpha^2)^{1/2}]/2$. This form corresponds to uniform marginals and increases the maximum correlation from $\frac{1}{3}$ to 0.4343. Similarly, for normal marginal distributions, the maximum correlation increases from 0.3183 to 0.4147. Huang and Kotz' generalization could be generated by the rejection method using Morgenstern's distribution as a dominating function.

The extension of Morgenstern's distribution to arbitrary dimensions has been accomplished in a set of papers by Johnson and Kotz (1975, 1977) and

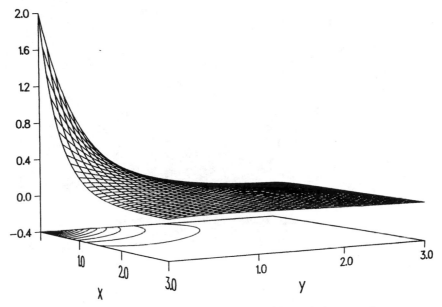

Figure 10.10. Morgenstern density, exponential marginals, $\alpha = +1$.

Cambanis (1977). Johnson and Kotz examine the n-dimensional distribution with density function:

$$f(u_1, u_2, \ldots, u_n) = 1 + \sum_{j_1}^{n-1} \sum_{<j_2}^{n} \beta_{j_1 j_2} \left[1 - 2u_{j_1}\right]\left[1 - 2u_{j_2}\right]$$

$$+ \sum_{j_1}^{n-2} \sum_{<j_2}^{n-1} \sum_{<j_3}^{n} \beta_{j_1 j_2 j_3}\left[1 - 2u_{j_1}\right]\left[1 - 2u_{j_2}\right]\left[1 - 2u_{j_3}\right]$$

$$+ \cdots + \beta_{12\ldots n} \prod_{j=1}^{n}\left(1 - 2u_j\right),$$

$$0 \leqslant u_j \leqslant 1, \; j = 1, 2, \ldots, n. \quad (10.3)$$

Every bivariate pair (U_i, U_j) has (marginal) distribution (10.1) with $\alpha = \beta_{ij}$. For large n, dealing with this multivariate distribution becomes something of a bookkeeping chore. The conditions on the β parameters so that (10.3) is indeed a legitimate density, however, are straightforward. Consider the 2^n

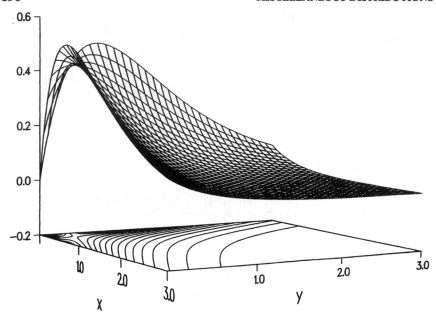

Figure 10.11. Morgenstern density, exponential marginals, $\alpha = -1$.

cases for $U_i = 0$ or 1, $i = 1, 2, \ldots, n$, and verify that $f(U_1, \ldots, U_n) \geq 0$. Of course, for $n = 2$, the condition is merely $|\beta_{12}| \leq 1$. For $n = 3$, the conditions can be conveniently summarized as follows:

$$|\beta_{13} + \beta_{23} \pm \beta_{123}| \leq 1 + \beta_{12}$$

$$|\beta_{12} + \beta_{23} \pm \beta_{123}| \leq 1 + \beta_{13}$$

$$|\beta_{12} + \beta_{13} \pm \beta_{123}| \leq 1 + \beta_{23}$$

$$|\beta_{123}| \leq 1 + \beta_{12} + \beta_{13} + \beta_{23}. \tag{10.4}$$

If in (10.4) the values $\beta_{12} = \beta_{13} = \beta_{23} = 1$ are sought, these inequalities force $\beta_{123} = 0$. Similarly if $\beta_{123} = 1$, then $\beta_{12} = \beta_{13} = \beta_{123} = 0$. Thus introducing "higher order" parameters serves to constrict the range of those β's with fewer subscripts. The primary use of Morgenstern's distribution would therefore appear to be in low-dimensional problems. For readers interested in other aspects of Morgenstern's distribution, the papers by Gupta and Wong (1984), Johnson (1982), and Lai (1978) may be of interest.

10.2. PLACKETT'S DISTRIBUTION

Plackett's (1965) bivariate uniform distribution is given by the density function

$$f(u, v) = \frac{\psi\{(\psi - 1)(u + v - 2uv) + 1\}}{\{[1 + (u + v)(\psi - 1)]^2 - 4\psi(\psi - 1)uv\}^{3/2}},$$

$$0 < u, v < 1, \psi > 0. \quad (10.5)$$

As indicated by the name, the marginal distributions are uniform 0–1. The parameter ψ governs the dependence between the components of (U, V) distributed according to f. Three cases indicate the role of ψ explicitly:

$$\psi \to 0 \qquad U = 1 - V$$
$$\psi = 1 \qquad U \text{ and } V \text{ are independent}$$
$$\psi \to \infty \qquad U = V$$

The parameter ψ has additional significance when examined in light of Plackett's original construction. Some aspects of this construction have been further honed by Mardia (1967, 1969) and Mosteller (1968). Let $H(x, y)$ denote a bivariate distribution function with marginal distribution functions $F(x)$ and $G(y)$. Consider an arbitrary point (x, y) in the plane and the four regions obtained by the two lines passing through (x, y) and parallel to one of the axes (Figure 10.12). The probability contents of these four regions are also indicated in Figure 10.12. Imagine now that the values p_1, p_2, p_3, and p_4 reside in a 2 × 2 table, as follows:

x

	p_2	p_4
y	p_1	p_3

The cross product ratio,

$$\psi = \frac{p_1 p_4}{p_2 p_3}, \quad (10.6)$$

measures the association between the two variables categorized in the table. Treating the probability contents in Figure 10.14 as probabilities and

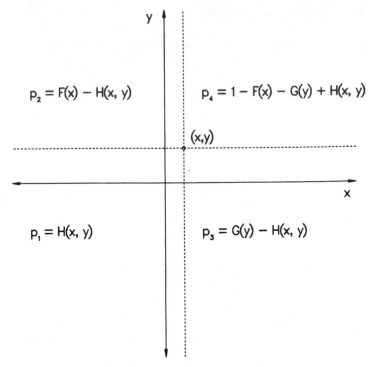

Figure 10.12. Construction of Plackett's distribution.

substituting in (10.6) gives

$$\psi = \frac{H(x, y)[1 - F(x) - G(y) + H(x, y)]}{[F(x) - H(x, y)][G(y) - H(x, y)]}, \qquad (10.7)$$

which is the defining equation for Plackett's distribution.

Plackett's family of distributions is the class of bivariate distribution functions having marginal distributions given by distribution functions F and G and having ψ in (10.7) constant for all choices of (x, y). The functional form of H can be found explicitly as

$$H(x, y) = \frac{S(x, y) - [S^2(x, y) - 4\psi(\psi - 1)F(x)G(y)]^{1/2}}{2(\psi - 1)}, \qquad \psi \neq 1$$

$$= F(x)G(y), \qquad \qquad \psi = 1,$$

$$(10.8)$$

where $S(x, y) = 1 + (\psi - 1)[F(x) + G(y)]$.

This calculation is straightforward recognizing that (10.7) is quadratic in H. Agreement of (10.5) and (10.8) is also evident by substituting $F(x) = x$ and $G(y) = y$ in (10.8) and then differentiating H with respect to x and y.

Some representative densities from Plackett's family of distributions are presented in Figures 10.13–10.15 for uniform, normal, and exponential marginals and for 18 values of ψ. Perhaps the most intriguing set of plots corresponds to the exponential marginals case (Figure 10.15). Of particular interest is the behavior of the density function for negative dependence ($\psi < 1$). For $\psi \geqslant 0.5$ the mode (indicated by a plus sign) of the joint density is at $(0, 0)$. For $\psi < 0.5$, the mode is located on the positive diagonal. The possibly surprising aspect of this observation is that the marginal density functions have modes at zero regardless of ψ.

To generate variates having Plackett's distribution, the conditional distribution approach is pertinent. First, a pair (U, V) having density f in (10.5) is generated. Arbitrary marginal distribution functions are then obtained by setting $X = F^{-1}(U)$, $Y = G^{-1}(V)$ so that (X, Y) has distribution function H in (10.8). The details of the conditional distribution approach for generating Plackett's distribution are due to Mardia (1967) and are embodied in the following algorithm:

1. Generate U_1 and U_2 independent uniform 0–1 and set $U = U_1$.
2. Set $V = [B - (1 - 2U_2)D]/2A$, where
   $$a = U_2(1 - U_2)$$
   $$A = \psi + a(\psi - 1)^2$$
   $$B = 2a(U_1\psi^2 + 1 - U_1) + \psi(1 - 2a)$$
   $$D = \psi^{1/2}[\psi + 4aU_1(1 - U_1)(1 - \psi)^2]^{1/2}.$$

The pair (U, V) has density f in (10.5).

A few other features of Plackett's distribution are mentioned that have some bearing on its use in Monte Carlo studies. Plackett noted the considerable similarity of his distribution with normal marginals to the usual bivariate normal distribution. In particular, if the correlation parameter $\rho = -\cos[\pi\psi^{1/2}/(1 + \psi^{1/2})]$, then the agreement in the corresponding distribution functions is at worst to a few units in the third decimal place. Plackett's distribution is also in a sense "close" to Morgenstern's distribution of Section 10.2. Conway (1979) showed by setting $\alpha = \psi - 1$ and expanding the radical in (10.8) in a Taylor's series about 0, that

$$H(x, y) = F(x)G(y)\{1 + \alpha[1 - F(x)][1 - G(y)]\} + O(\alpha).$$

Especially in the vicinity of independence, Plackett's and Morgenstern's distributions approximate each other.

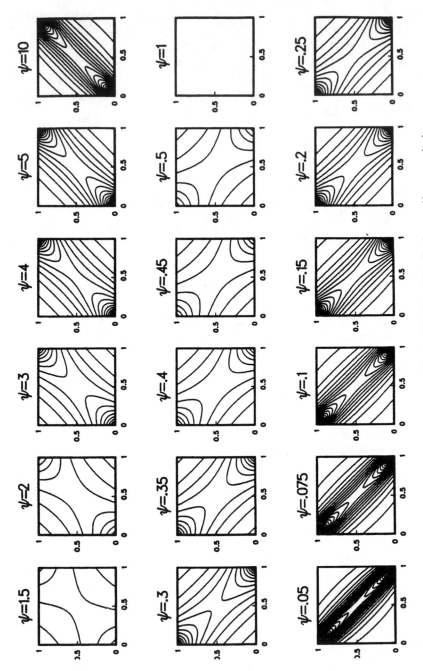

Figure 10.13. Contour plots for Plackett's family of distributions, uniform marginals.

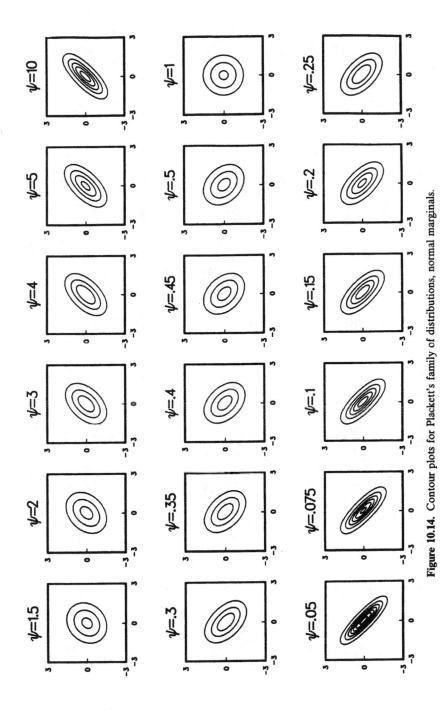

Figure 10.14. Contour plots for Plackett's family of distributions, normal marginals.

195

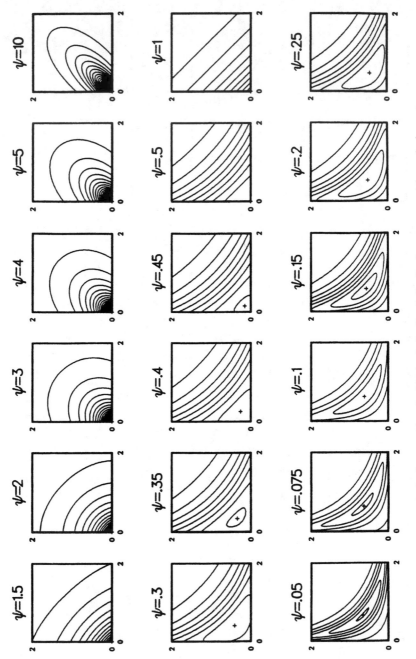

Figure 10.15. Contour plots for Plackett's family of distributions, exponential marginals.

196

In summary, Plackett's distribution has a number of pleasant characteristics including a simple density and distribution function, a straightforward generation algorithm, and a full range of dependence from completely negative ($U = 1 - V$) through independence to completely positive ($U = V$). On the minus side, the distribution is quite similar to two other distributions (the "usual" normal and Morgenstern's distribution). This implies, for example, that only small departures from the bivariate normal can be examined with Plackett's distribution having normal marginals. A final restriction with Plackett's distribution that was not previously noted is that it is strictly limited to the bivariate case. Conway (1979) noted a reasonable extension to three dimensions that is related to association for a $2 \times 2 \times 2$ contingency table. However, the results are not yet complete for this extension.

10.3. GUMBEL'S BIVARIATE EXPONENTIAL DISTRIBUTION

Gumbel (1961) studied the bivariate exponential distribution with density function

$$f(x, y; \theta) = [(1 + \theta x)(1 + \theta y) - \theta]\exp(-x - y - \theta xy),$$

$$x, y > 0; 0 \leqslant \theta \leqslant 1. \quad (10.9)$$

The marginal distributions are standard exponential, and if θ equals zero, then the components are independent. At the other extreme, the correlation with $\theta = 1$ is -0.43, which indicates a relatively weak negative dependence.

The conditional distribution approach is appropriate for deriving a simple generation algorithm. The conditional distribution of Y given $X = x$ has density function

$$f(y) = e^{-y-\theta xy}[(1 + \theta x)(1 + \theta y) - \theta], \qquad y > 0. \quad (10.10)$$

This density can be manipulated to become

$$f(y) = p[\beta e^{-\beta y}] + (1 - p)[\beta^2 y e^{-\beta y}],$$

where $\beta = 1 + \theta x$ and $p = (\beta - \theta)/\beta$. The bracketed terms are themselves density functions—the first is exponential and the second is a $\Gamma(2, \beta)$ density. Thus the conditional density f corresponds to a probabilistic mixture. The following algorithm can be used to generate (X_1, X_2)

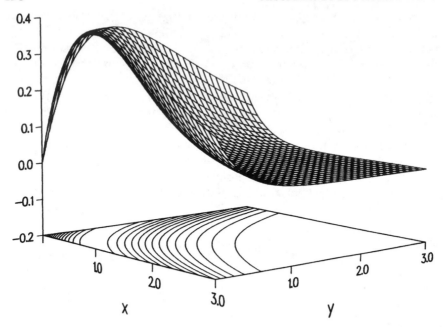

Figure 10.16. Gumbel exponential density, $\theta = +1$.

distributed as f in (10.9):

1. Generate U_1, U_2, and U_3, independent uniform 0–1.
2. Set $X_1 = -\ln(U_1)$.
3. Compute $\beta = 1 + \theta X_1$ and $p = (\beta - \theta)/\beta$.
4. Set $Y = -\ln(U_2)$.
5. If U_3 is less than p, set $X_2 = \beta Y$ and stop.
6. Otherwise, generate U_4 uniform 0–1, set $X_2 = \beta(Y - \ln U_4)$, and stop.

Figure 10.16 illustrates the Gumbel exponential density for θ equal to 1. For $\theta = 0$, the independence case, the density would be identical to that depicted in Figure 10.9. Transforming via $U = 1 - \exp(-X)$ and $V = 1 - \exp(-Y)$ yields uniform 0–1 components whose joint density is pictured in Figure 10.17. Further transforming to normal marginals according to $\Phi^{-1}(U)$, $\Phi^{-1}(V)$ yields the density given in Figure 10.18. The final plot for the Gumbel family (Figure 10.19) represents the density of the pair $\Phi^{-1}(U)$, $\Phi^{-1}(1 - V)$, thus rigged to obtain positive dependence between the components.

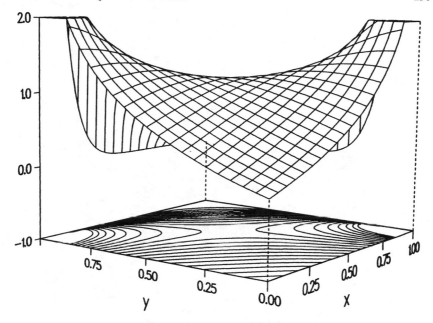

Figure 10.17. Gumbel uniform density, $\theta = +1$.

10.4. ALI-MIKHAIL-HAQ'S DISTRIBUTION

Ali, Mikhail, and Haq (1978) devised a bivariate uniform distribution with distribution function

$$F(u, v) = \frac{uv}{1 - \alpha(1 - u)(1 - v)}, \qquad 0 < u, v < 1; -1 \leqslant \alpha \leqslant 1. \quad (10.11)$$

For α equal to zero, U and V are independent. For α equal to one, the pair $[\ln(U/(1 - U)), \ln(V/(1 - V))]$ has Gumbel's (1961) Type I logistic distribution.

The key idea underlying the development (10.11) is related to an odds function. In the univariate case, the odds function is defined as

$$H(x) = \frac{1 - G(x)}{G(x)}, \qquad (10.12)$$

where G is a distribution function. If X is a random variable distributed as

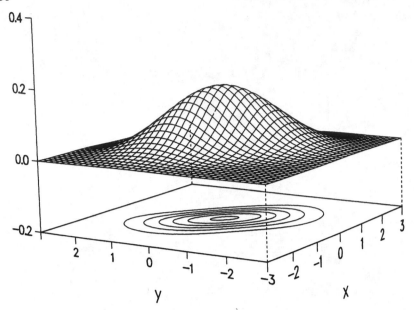

Figure 10.18. Gumbel normal density, $\theta = 1$.

G, then $H(x)$ represents the odds of X taking on a value larger than x versus a value of X smaller than x. Solving for G in (10.12) yields

$$G(x) = \frac{1}{1 + H(x)}. \tag{10.13}$$

A valid distribution function G is obtained in (10.13) if H is a nonincreasing function with values declining from $+\infty$ to 0 across the range of x.

A natural extension to the bivariate case is to consider the defining equation

$$G(x, y) = \frac{1}{1 + H(x, y)},$$

where H is a bivariate odds function with marginal odds functions H_1 and H_2. The conditions on H so that G is a distribution function are given by Ali, Mikhail, and Haq as

1. $H(x, \infty) = H_1(x)$, $H(\infty, y) = H_2(y)$, $H(-\infty, y) = H(x, -\infty)$
$= \infty$.

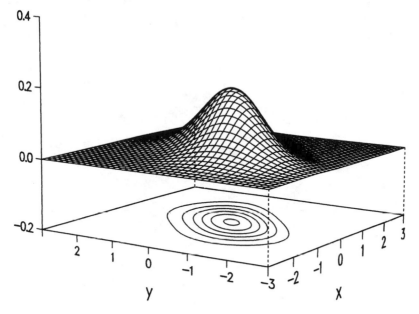

Figure 10.19. Gumbel normal density, $\theta = 1$.

2. For $x_1 \leqslant x_2$ and $y_1 \leqslant y_2$,
$$[1 + H(x_2, y_2)]^{-1} + [1 + H(x_1, y_1)]^{-1} - [1 + H(x_1, y_2)]^{-1}$$
$$- [1 + H(x_2, y_1)]^{-1} \geqslant 0.$$

The function H defined by

$$H(x, y) = H_1(x) + H_2(y) + (1 - \alpha)H_1(x)H_2(y), \qquad -1 \leqslant \infty \leqslant 1,$$

satisfies these conditions and determines the distribution function

$$G(x, y) = \frac{G_1(x)G_2(y)}{1 - \alpha[1 - G_1(x)][1 - G_2(y)]}, \qquad (10.14)$$

where G_1 and G_2 denote the marginal distribution functions. Substituting $G_1(x) = x$ and $G_2(y) = y$ yields the form given initially in (10.11).

Generation of variates (U, V) from (10.11) is accomplished using the conditional distribution approach. The general form in (10.14) is then realized using $X = G_1^{-1}(U)$ and Y and $G_2^{-1}(V)$. The conditional distribu-

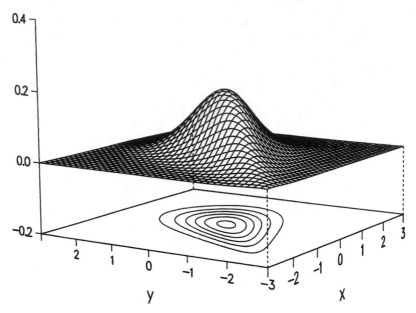

Figure 10.20. Ali-Mikhail-Haq density, normal marginals, $\alpha = 1$.

tion of V given $U = u$ in (10.11) is

$$F(v|u) = \frac{v[1 - \alpha(1 - v)]}{[1 - \alpha(1 - u)(1 - v)]^2}.$$

The equation $F(v|u) = U_2$ where U_2 is uniform 0–1 is quadratic in v. Generation is straightforward and is accomplished by the following algorithm:

1. Generate U_1 and U_2, independent uniform 0–1, and set $U = U_1$.
2. Compute the following:
   $b = 1 - U_1$
   $A = -\alpha(2bU_2 + 1) + 2\alpha^2 b^2 V_2 + 1$
   $B = \alpha^2(4b^2 U_2 - 4bU_2 + 1) + \alpha(4bU_2 - 4p + 2) + 1.$
3. Set $V = 2V_2(\alpha b - 1)^2/(A + \sqrt{B})$.

For comparison, two normal marginals plots are given in Figures 10.20 and 10.21 It should be somewhat apparent that even for the extreme value of $\alpha = 1$, the dependence is not particularly strong. Conway (1979) investigated this point and found, for example, that the Pearson correlation in (10.11) is restricted to the range -0.271 to 0.478 and in (10.14) for

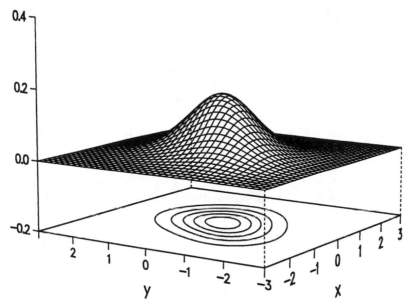

Figure 10.21. Ali-Mikhail-Haq density, normal marginals, $\alpha = 0.5$.

exponential marginals to the range -0.227 to 0.290. For normal marginals, the range is approximately -0.300 to 0.600.

Conway also noted an interesting relationship between Ali-Mikhail-Haq and Morgenstern's distribution. The distribution function (10.11) can be expanded as

$$F(u, v) = uv\left[1 + \alpha(1 - u)(1 - v)\right] + \sum_{i=2}^{\infty} \alpha^i(1 - u)^i(1 - v)^i. \quad (10.15)$$

The first term is Morgenstern's distribution function for uniform marginals. Huang and Kotz (1984) have shown the expression in (10.15) remains a valid distribution function if the series is truncated. The range of correlations under this model is only slightly enlarged, however.

Finally, for $\alpha = 1$, the Ali-Mikhail-Haq distribution is identical to the uniform marginals case of the Burr-Pareto-logistic form given in Section 9.1.

10.5. WISHART DISTRIBUTION

The Wishart distribution is one of the few multivariate distributions for which computer generation algorithms have been published. Outside the normal distribution theory framework, however, the Wishart distribution has limited appeal. Hence, the treatment in this section focuses on variate generation.

A $p \times p$ symmetric matrix A has a Wishart distribution with parameters Σ, n, and p and denoted $W_p(n, \Sigma)$, if it has the same distribution as $\sum_{i=1}^{n} \mathbf{X}_i \mathbf{X}_i'$, where the $\mathbf{X}_i$'s are independent and identically distributed as $N_p(\mathbf{0}, \Sigma)$. This constructive definition immediately gives rise to a generation algorithm: generate n multivariate normal random vectors and form $\sum_{i=1}^{n} \mathbf{X}_i \mathbf{X}_i'$. This approach requires a total of np univariate normal variates to form $p(p + 1)/2$ distinct entries in the $W_p(n, \Sigma)$ distribution. For this reason, this direct but computationally expensive method is enthusiastically ridiculed by developers of alternate algorithms. For non-normal distributions, however, generation of the sample covariance matrix would seem to require this crude but direct approach.

An obvious method for generating the Wishart distribution is to appeal to Bartlett's (1933) decomposition. Let T be a lower triangular $p \times p$ matrix with entries T_{ij} satisfying:

1. T_{ij} is standard normal for $i > j$.
2. T_{ii} is $\sqrt{\chi^2_{(n-i+1)}}$, for $i = 1, \ldots, p$.
3. The T_{ij}'s are independent.

The matrix $A = TT'$ has a $W_p(n, I)$ distribution. Handling a matrix other than the identity matrix is easy. If the $p \times p$ symmetric matrix Σ has the Choleski factorization $\Sigma = LL'$, with L lower triangular, then $B = LAL'$ has a $W_p(n, \Sigma)$ distribution.

A number of papers have appeared addressing the problem of Wishart variate generation. Odell and Feiveson (1966) basically give the above algorithm, although they credit it to Hartley and Harris (1963) rather than Bartlett. Smith and Hocking (1972) developed a computer program to implement the method using the Wilson-Hilferty approximation to handle the generation of the diagonal elements. At the time, that was a convenient albeit approximate method to accomplish the task. As Section 2.5 attests, there are now many exact methods to generate gamma variates. Recently, Jones (1985) adapted the Smith-Hocking program to produce inverse Wishart variates. Some other problems related to Wishart generation have been considered. Gleser (1976) and Johnson and Hegemann (1974) developed methods for generating the noncentral Wishart distribution. Marasinghe and Kennedy (1982) have devised schemes for generating extreme characteristic roots of the Wishart distribution and have noted the interest of nuclear physicists for this problem.

CHAPTER 11

Research Directions

A significant undercurrent in previous chapters was the interplay of variate generation algorithms and schemes for constructing distributions. Construction schemes invariably led directly to variate generation algorithms, and algorithms (bared of the computational frills to get high efficiency) at times revealed a basic construction. In this chapter some of the key construction devices are highlighted—thus serving as a commentary review of the previous material with a slightly different emphasis. More importantly, this review provides a framework for describing other extensions to these distributions and some potentially fruitful avenues of research.

One of the easiest pathways to constructing distributions is to apply transformations to existing distributions. This was the device in constructing Johnson's translation system (Chapter 5) and the resulting family has had an enviable track record indeed. Specific directions of pursuit involve two choices: the family of transformations and the original distribution to be transformed. Here are a few possibilities:

1. Begin with the multivariate normal distribution and apply Tukey's g and h transformations to the components to give rise to a multivariate g-h family of distributions. The g-h transformation to a normal variate Z is

$$X = g^{-1}[\exp(gZ) - 1]\exp\left(\frac{hZ^2}{2}\right).$$ (11.1)

For g and h both zero, the normal distribution is retained and for h equal to zero, the lognormal family of distributions occurs. The parameter g effects the symmetry of the distribution and h impacts the heaviness of the tails. The g-h transformation has been examined in the literature by Hoaglin and Peters (1979) and by Martinez and Iglewicz (1984). It may also be worth trying this transformation (11.1) on other multivariate distributions with normal component distributions. For analytical ease, the multivariate normal distribution is suggested as a first step.

2. Start with multivariate distributions with uniform marginal distributions and transform via the lambda distribution (Section 2.1). Johnson and Kotz (1973) examined this tack when the starting point was a joint Dirichlet distribution. Obviously, other multivariate distributions with uniform marginals could be tried.

3. Another approach is to use the Johnson transformations (identity, exponential, hyperbolic sine, and inverse logit) applied to the multivariate logistic distribution. The univariate groundwork is already in place (Tadikamalla and Johnson, 1982a, b).

The above examples represent specific constructions and are obvious extensions to existing work (the mathematical details in exploring the properties of these distributions are not trivial, however). A more general set of possibilities involves identifying useful transformations and then perturbing them in an interesting fashion. Since this idea was exploited frequently in earlier chapters, some of these examples are now reviewed in this framework.

Virtually anyone with a slight exposure to mathematical statistics knows that the square of a standard normal variate is chi-square with one degree of freedom. This relationship can be inverted to give

$$Z = \pm\sqrt{X} \tag{11.2}$$

where X is $\chi^2_{(1)}$ and "$\pm$" represents a random sign. Equation (11.2) is not generally used in normal variate generation since better methods are available (Section 2.2). The merit of this approach will become apparent in extending (11.2) to two dimensions. Let $\mathbf{B} = (B_1, B_2)$ be uniform on the unit circle and let X be $\chi^2_{(2)}$ (independent of $\mathbf{B}$). Form the independent normal variates

$$Z_1 = B_1\sqrt{X}$$

$$Z_2 = B_2\sqrt{X}. \tag{11.3}$$

For generation purposes, it is useful to rewrite this in a slightly different form. Since X is $\chi^2_{(2)}$ it is also exponential so that it can be generated as $-2\ln U_1$. Also, $\mathbf{B}$ can be represented as $(\cos\theta, \sin\theta)$ with θ uniform on

$(0, 2\pi)$. Applying these results to (11.3), we get

$$Z_1 = \cos(2\pi U_2)(-2\ln U_1)^{1/2}$$

$$Z_2 = \sin(2\pi U_2)(-2\ln U_1)^{1/2}, \tag{11.4}$$

which is the Box-Muller transformation (Section 2.2). As noted previously, the polar representation is available in higher dimensions (Section 7.1).

With the Box-Muller transformation (11.4) stripped to its bare essentials in (11.3), obvious generalizations emerge. First, if X has other than the proper chi-square distributions, what Z distribution results? Of course, the answer is the spherical distributions given in Chapter 6. There it was noted that selecting X to be Pearson Type VI or beta gave a useful coverage to this class of distributions. The negative implication is that there is little room for useful generalizations by varying the distribution of X.

On the other hand, Section 7.2 pursued a line of investigation concerned with varying the distribution of **B** from the baseline uniform. This approach gave a number of useful new distributions. Generally, that strategy retained X as $\chi^2_{(n)}$ (to facilitate comparisons with the baseline normal model), although this could be altered. Another fruitful avenue is to introduce dependence between **B** and X. For example, suppose U_1 and U_2 in (11.4) have Morgenstern's uniform distribution (Section 10.1).

A direct way of introducing dependence between the direction **B** and the multiplier X is to specify the conditional distribution of X given $\mathbf{B} = \mathbf{b}$. Suppose, for example, the univariate direction Θ has the density function $f(\theta) = (1 + \sin 2\theta)^{-1}$, $0 < \theta < \pi/2$. Given $\Theta = \theta$, let X be gamma distributed with shape parameter 2 and scale parameter $\cos\theta + \sin\theta$. If we form

$$X_1 = X\cos\theta$$

$$X_2 = X\sin\theta,$$

the resulting distribution is bivariate exponential with independent components. This example suggests that other dependence relationships between Θ and X might also be worth investigating.

Another way of extending (11.3) is to alter the support of **B**. In other words, what happens if instead of being on the "surface" of the unit interval (at -1 or $+1$) or the boundary of the unit circle, the support is on the interior of the corresponding region? Johnson (1976) showed that to get normal variates out of this relationship (11.3), the multiplier chi-square random variable needs two more degrees of freedom. In one dimension, this means that a normal variate can be represented as $X = ZU$, where U is uniform on the full interval $(-1, 1)$ and Z is $\sqrt{\chi^2_{(3)}}$. Attaching the random sign to Z rather than U gives

$$X = \pm\sqrt{\chi^2_{(3)}}\, U, \tag{11.5}$$

where U is uniform 0–1. This equation is what we should expect upon consideration of Khintchine's unimodality theorem (Chapter 8). If we consider other powers besides square root and if we let the degrees of freedom vary, then the generalized exponential power distribution (Section 2.4) is obtained. Chapter 8 was concerned with multivariate and especially bivariate extensions of (11.5).

Pursuing the lines of interesting perturbations eventually led to the "well-known" Khintchine theorem, which provides a unifying framework for many of the distributions considered previously. Other extensions may be possible by considering the following papers, which are briefly reviewed. Here again, if a kernel transformation can be identified, then extensions become obvious. Olshen and Savage (1970) examined unimodality for multivariate distributions. In light of this Wolfe (1975) looked at spherically symmetric stable distributions. The strategy of perturbing the characteristic function of the multivariate normal is explored by Silva (1978) and Cambanis, Keener, and Simons (1983). Cambanis and colleagues give some specific constructions such as

$$X = RX_0$$

$$Y = RY_0,$$

where R is nonnegative and $(X_0, Y_0) = [UB^{-1/2}, V(1 - B)^{-1/2}]$, with (U, V) uniform on the unit circle and independently, B is beta $(\frac{1}{2}, \frac{1}{2})$. Of other possible interest are the reports by Bertin and Theodorescu (1980, 1983), which give more mathematical descriptions of unimodality. At least, in the univariate realm, algorithms can be developed for generating variates with unimodal densities (Marsaglia and Tsang, 1984 and Devroye, 1984). Perhaps these approaches can be extended to higher dimensions.

Various distributions considered in earlier chapters have fairly direct (in theory) extensions. It may be apparent, for example, that the Burr-Pareto-logistic family merged with Morgenstern's distribution (Section 9.2) could use instead other multivariate exponential distributions. Some possibilities are reviewed by Block and Savits (1981). Methods for sampling from some of these distributions have been given by Friday (1976). Extensions to other distributions were noted at the time for the Morgenstern (Section 10.1), Plackett (10.2), and Ali (Section 10.4) distributions.

Some emphasis has been placed in earlier chapters on multivariate distributions having uniform marginal distributions. These distributions are convenient since any specified component distribution can be obtained by the inverse probability transformation. In effect, the specification of component distributions becomes separated from the intrinsic dependence structure. This point was probably most apparent with the Burr-Pareto-logistic distributions in Chapter 9.

Some related papers fall under the heading of distributions with specified marginals. This was the motivating force behind the distributions derived by Johnson and Tenenbein (1981). They looked at the bivariate constructions

$$Y_1 = X_1$$

$$Y_2 = cX_1 + (1 - c)X_2 \text{ (weighted linear combination)}$$

and

$$Y_1 = X_1 + \beta X_3$$

$$Y_2 = X_2 + \beta X_3 \text{ (trivariate reduction)},$$

where the X_i's were uniform, Laplace, or exponential. The algebra even in this bivariate case with these simple choices of X_i was horrendous and the authors made use of MACSYMA, a computer algebra code, to accomplish the derivations. In short, they demonstrated the difficulty in obtaining useful classes of distributions having specified marginal distributions.

Without going into specific parametric constructions, Whitt (1976) was able to obtain general results on the range of possible product moment correlations in certain families. These results were put to good use by Schmeiser and Lal (1982), who looked at constructions such as

$$Y_1 = F_1^{-1}(U) + X_1 + X_3$$

$$Y_2 = F_2^{-1}(V) + X_2 + X_3,$$

where the X_i's are $\Gamma(\alpha_i, 1)$ independent of the uniform variate U, $V = U$ or $1 - U$ (prespecified), and the F_i's are gamma distribution functions.

Some more general (mathematical) treatments of the specified marginals problem have been given by Strassen (1965) originally, with later efforts by Heilmann (1981) and Gaffke and Rüschendorf (1984).

Of related interest are the many time-series type models examined by Lawrance and Lewis (1983). These distributions have some desirable characteristics—they are easy to generate and have been known to emulate real data. One of their more general constructions (which includes many useful special cases) has the form

$$X_1 = \beta_1 V_1 E_1 + I_1 E_2$$

$$X_2 = \beta_2 V_2 E_2 + I_2 E_1,$$

where E_1 and E_2 are independent standard exponential variates, the V_i's and I_i's are binary variables defined by

$$V_i = 1 \qquad \text{with probability } \alpha_i$$

$$= 0 \qquad \text{with probability } 1 - \alpha_i,$$

$$I_i = 1 \qquad\qquad \text{with probability } \frac{1 - \beta_i}{1 - (1 - \alpha_i)\beta_i}$$

$$= (1 - \alpha_i)\beta_i \qquad \text{with probability } \frac{\alpha_i \beta_i}{1 - (1 - \alpha_i)\beta_i}.$$

It is assumed that the pairs (E_1, E_2), (V_1, V_2), and (I_1, I_2) are independent, although within a pair, the variables may be dependent. In particular, the correlation between X_1 and X_2 is

$$\rho(X_1, X_2) = \beta_1 \beta_2 \text{Cov}(V_1, V_2) + \text{Cov}(I_1, I_2)$$

$$+ \alpha_1 \beta_1 (1 - \alpha_2 \beta_2) + \alpha_1 \beta_1 (1 - \alpha_2 \beta_2).$$

The exponential autoregressive form of this general model has been elegantly extended by Raftery (1982).

The discussion of this chapter is in no way intended to be delimiting. In fact it is hoped that other schemes could be suggested. In perusing journals for such ideas, there is ample opportunity to discover results leading to specific constructions of useful multivariate distributions for simulation applications.

References

Ahrens, J. H. and U. Dieter (1974). "Computer Methods for Sampling from Gamma, Beta, Poisson, and Binomial Distributions," *Computing*, **12**, 223–246.

Ali, M. M., N. N. Mikhail, and M. S. Haq (1978). "A Class of Bivariate Distributions Including the Bivariate Logistic," *Journal of Multivariate Analysis*, **8**, 405–412.

Anderson, T. W. (1966). "Some Nonparametric Procedures Based on Statistically Equivalent Blocks," *Proceedings of an International Symposium on Multivariate Analysis* (P. R. Krishnaiah, Ed.). Academic, New York, 5–27.

Atkinson, A. C. (1982), "The Simulation of Generalized Inverse Gaussian and Hyperbolic Random Variables," *SIAM Journal of Scientific and Statistical Computation*, **3**, 502–515.

Bartels, R. (1982). "The Rank Version of von Neumann's Ratio Test for Randomness," *Journal of the American Statistical Association*, **77**, 40–46.

Bartlett, M. S. (1933). "On the Theory of Statistical Regression," *Proceedings of the Royal Society of Edinburgh*, **53**, 260–283.

Beauchamp, J. J. and V. E. Kane (1982). "Robustness of Three Power Transformation Estimation Procedures," *Journal of Statistical Computation and Simulation*, **15**, 251–272.

Beckman, R. J. and R. D. Cook (1983), "Outlier...s," *Technometrics*, **25**, 119–163.

Beckman, R. J. and M. E. Johnson (1981). "A Ranking Procedure for Partial Discriminant Analysis," *Journal of the American Statistical Association*, **76**, 671–675.

Bertin, E. M. J. and R. Theodorescu (1980). "Some Characterizations of Discrete Unimodality," University of Utrecht Technical Report.

Bertin, E. M. J and R. Theodorescu (1983). "Hincin Spaces and Unimodal Probability Measures," University of Utrecht Technical Report.

Bingham, C. (1974). "An Antipodally Symmetric Distribution on the Sphere," *The Annals of Statistics*, **2**, 1201–1225.

Block, H. W. and T. H. Savits (1981). "Multivariate Distributions in Reliability Theory and Life Testing," *Statistical Distributions in Scientific Work*, **5**, 271–288.

Borwein, J. M. and P. B. Borwein (1984). "The Arithmetic-Geometric Mean and Fast Computation of Elementary Functions," *SIAM Review*, **26**, 351–370.

Box, G. E. P. and M. E. Muller (1958). "A Note on the Generation of Random Normal Deviates," *The Annals of Mathematical Statistics*, **29**, 610–611.

Box, G. E. P. and G. C. Tiao (1973). *Bayesian Inference in Statistical Analysis*. Addison-Wesley, Reading, Massachusetts.

Bratley, P., B. L. Fox, and L. E. Schrage (1983). *A Guide to Simulation*. Springer-Verlag, New York.

Broffitt, J. D., R. H. Randles, and R. V. Hogg (1976). "Distribution-Free Partial Discriminant Analysis," *Journal of the American Statistical Association*, **71**, 934–939.

Bryson, M. C. and M. E. Johnson (1982). "Constructing and Simulating Multivariate Distributions Using Khintchine's Theorem," *Journal of Statistical Computation and Simulation*, **16**, 129–137.

Burr, I. W. (1942). "Cumulative Frequency Functions," *Annals of Mathematical Statistics*, **13**, 215–232.

Cambanis, S. (1977). "Some Properties and Generalizations of Multivariate Eyraud-Gumbel-Morgenstern Distributions," *Journal of Multivariate Analysis*, **7**, 551–559.

Cambanis, S., S. Huang, and G. Simons (1981). "On the Theory of Elliptically Contoured Distributions," *Journal of Multivariate Analysis*, **11**, 368–385.

Cambanis, S., R. Keener, and G. Simons (1983). "On α-Symmetric Multivariate Distributions," *Journal of Multivariate Analysis*, **13**, 213–233.

Chambers, J. M., C. L. Mallows, and B. W. Stuck (1976). "A Method for Simulating Stable Random Variables," *Journal of the American Statistical Association*, **71**, 340–344.

Cheng, R. C. H. and G. M. Feast (1979). "Some Simple Gamma Variate Generators," *Applied Statistics*, **28**, 290–295.

Chmielewski, M. A. (1981). "Elliptically Symmetric Distributions: A Bibliography and Review," *International Statistical Review*, **49**, 67–74.

Chu, K. (1973). "Estimation and Decision for Linear Systems with Elliptical Random Processes," *IEEE Transactions on Automatic Control*, **18**, 499–505.

Conover, W. J. and R. L. Iman (1980). "The Rank Transformation as a Method of Discrimination with Some Examples," *Communications in Statistics*, **A9**, 465–487.

Conway, D. A. (1979). "Multivariate Distributions with Specified Marginals," Stanford University Technical Report.

Cook, J. M. (1957). "Rational Formulae for the Production of a Spherically Symmetric Probability Distribution," *Mathematics of Computation*, **11**, 81–82.

Cook, R. D. and M. E. Johnson (1981). "A Family of Distributions for Modelling Non-Elliptically Symmetric Multivariate Data," *Journal of the Royal Statistical Society*, Series B, **43**, 210–218.

Cook, R. D. and M. E. Johnson (1986). "Generalized Burr-Pareto-Logistic Distributions with Applications to a Uranium Exploration Data Set," *Technometrics*, **28**, 123–131.

Das Gupta, S., M. L. Eaton, I. Olkin, M. Perlman, L. J. Savage, and M. Sobel (1972). "Inequalities on the Probability Content of Convex Regions for Elliptically Contoured Distributions," *Proceedings of the Sixth Berkeley Symposium on Mathematical Statistics and Probability*, **2**, 241–265.

Devlin, S. J., R. Gnanadesikan, and J. R. Kettenring (1976). "Some Multivariate Applications of Elliptical Distributions," *Essays in Probability and Statistics* (S. Ikeda et al., Eds.). Shinko Tsusho Co. Ltd., Tokyo.

Devroye, L. (1981). "The Series Method for Random Variate Generation and its Application to the Kolmogorov-Smirnov Distribution," *American Journal of Mathematical and Management Sciences*, **1**, 359–379.

Devroye, L. (1984). "Random Variate Generation for Unimodal and Monotone Densities," *Computing*, **32**, 43–68.

Dongarra, J. J., C. B. Moler, J. R. Bunch, and G. W. Stewart (1979). *LINPACK User's Guide*, SIAM, Philadelphia.

Dudewicz, E. J. and T. G. Ralley (1981). *The Handbook of Random Number Generation and Testing with TESTRAND Computer Code*. American Sciences Press, Columbus, Ohio.

Durling, F. C. (1975). "The Bivariate Burr Distribution," *Statistical Distributions in Scientific Work*, **1**, 329–335.

Efron, B. (1979). "Bootstrap Methods: Another Look at the Jackknife," *Annals of Statistics*, **7**, 1–26.

Eisenberger, I. (1964). "Genesis of Bimodal Distributions," *Technometrics*, **6**, 357–363.

Everitt, B. S. (1979). "A Monte Carlo Investigation of the Robustness of Hotelling's One- and Two-Sample T^2 Tests," *Journal of the American Statistical Association*, **74**, 48–51.

Eyraud, H. (1938). "Les Principes de la Mesure des Correlations," *Annales Universite Lyons*, *Series A, Sciences Mathematiques et Astonomie*, **1**, 30–47.

Farlie, D. J. G. (1960). "The Performance of Some Correlation Coefficients for a General Bivariate Distribution," *Biometrika*, **47**, 307–323.

Feller, W. (1971). *An Introduction to Probability Theory and Its Applications*, Vol. II, 2nd ed. John Wiley & Sons, New York.

Foutz, R. V. (1980). "A Test for Goodness-of-Fit Based on an Empirical Probability Measure," *Annals of Statistics*, **8**, 989–1001.

Franke, R. and T. Jayachandran (1983). "An Empirical Investigation of the Properties of a New Goodness of Fit Test," *Journal of Statistical Computation and Simulation*, **16**, 175–187.

Freund, R. J. (1961). "A Bivariate Extension of the Exponential Power Distribution," *Journal of the American Statistical Association*, **56**, 971–977.

Friday, D. E. (1976). "Computer Generation of Random Numbers from Bivariate Life Distributions," *Proceedings of the Ninth Interface Symposium on Computer Science and Statistics*, 218–221.

Fushimi, M. and S. Tezuka, (1983). "The k-Distribution of Generalized Feedback Shift Register Pseudorandom Numbers," *Communications of the ACM*, **26**, 516–523.

Gaffke, N. and L. Rüschendorf (1984). "On the Existence of Probability Measures with Given Marginals," *Statistics and Decisions*, **2**, 163–174.

Gleser, L. J. (1976). "A Canonical Representation for the Noncentral Wishart Distribution Useful for Simulation," *Journal of the American Statistical Association*, **71**, 690–695.

Gumbel, E. J. (1958). "Distributions a Plusiers Variables dont les Marges sont Donnees," *Comptes Rendus*, **246**, 2717–2719.

Gumbel, E. J. (1960). "Bivariate Exponential Distributions," *Journal of the American Statistical Association*, **55**, 698–707.

Gumbel, E. J. (1961). "Bivariate Logistic Distributions," *Journal of the American Statistical Association*, **56**, 335–349.

Gupta, A. K. and C. F. Wong (1984). "On a Morgenstern-Type Bivariate Gamma Distribution," *Metrika*, **31**, 327–332.

Hartley, H. O. and D. L. Harris (1963). "Monte Carlo Computations in Normal Correlation Problems," *Journal of Association for Computing Machinery*, **10**, 301–306.

Hawkins, D. M. (1981). "A New Test for Multivariate Normality and Homoscedasticity," *Technometrics*, **23**, 105–110.

Heilmann, W.-R. (1981). "A Mathematical Programming Approach to Strassen's Theorem on Distributions with Given Marginals," *Mathematische Operationsforschung und Statistik, Series Optimization*, **12**, 593–596.

Herz, C. S. (1955). "Bessel Functions of Matrix Argument," *Annals of Mathematics*, **61**, 474–523.

Hicks, J. S. and R. F. Wheeling (1959). "An Efficient Method for Generating Uniformly Distributed Points on the Surface of an *n*-Dimensional Sphere," *Communications of the ACM*, **2**, 17–19.

Higgins, J. J. (1975). "A Generalized Method of Constructing Multivariate Densities and Some Related Inferential Procedures," *Communications in Statistics*, **4**, 955–966.

Hoaglin, D. C. and S. C. Peters (1979). "Software for Exploring Distribution Shape," *Proceedings of the Waterloo Interface of Statistics and Computer Science*.

Huang, J. S. and S. Kotz (1984). "Correlation Structure in Iterated Farlie-Gumbel-Morgenstern Distributions," *Biometrika*, **71**, 633–636.

Hutchinson, T. P. (1979). "Four Applications of a Bivariate Pareto Distribution," *Biometrical Journal*, **21**, 553–563.

IMSL Package (1980). IMSL, Inc., Houston, Texas.

Jensen, D. R. (1970). "A Generalization of the Multivariate Rayleigh Distribution," *Sankhya*, Series A, **32**, 193–208.

Johnson, D. E. and V. Hegemann (1974). "Procedures to Generate Random Matrices with Noncentral Distributions," *Communications in Statistics*, **3**, 691–699.

Johnson, M. E. (1976). "Models and Methods for Generating Dependent Random Vectors," Ph.D. Dissertation, University of Iowa.

Johnson, M. E. (1979). "Computer Generation of the Exponential Power Distribution," *Journal of Statistical Computation and Simulation*, **9**, 239–240.

Johnson, M. E. (1982). "Computer Generation of the Generalized Eyraud Distribution," *Journal of Statistical Computation and Simulation*, **15**, 333–335.

Johnson, M. E., J. S. Ramberg, and C. Wang (1982). "The Johnson Translation System in Monte Carlo Studies," *Communications in Statistics—Computation and Simulation*, **11**, 521–525.

Johnson, M. E. and A. Tenenbein (1981). "A Bivariate Distribution Family with Specified Marginals," *Journal of the American Statistical Association*, **76**, 198–201.

Johnson, M. E., G. L. Tietjen, and R. J. Beckman (1980). "A New Family of Probability Distributions with Applications to Monte Carlo Studies," *Journal of the American Statistical Association*, **75**, 276–279.

Johnson, N. L. (1949a). "Bivariate Distributions Based on Simple Translation Systems," *Biometrika*, **36**, 297–304.

Johnson, N. L. (1949b). "Systems of Frequency Curves Generated by Methods of Translation," *Biometrika*, **36**, 149–176.

Johnson, N. L. and S. Kotz (1969). *Distributions in Statistics: Discrete Distributions*. John Wiley & Sons, New York.

Johnson, N. L. and S. Kotz (1972). *Distributions in Statistics: Continuous Multivariate Distributions*. John Wiley & Sons, New York.

Johnson, N. L. and S. Kotz (1973). "Extended and Multivariate Tukey Lambda Distributions," *Biometrika*, **60**, 655–661.

Johnson, N. L. and S. Kotz (1975). "On Some Generalized Farlie-Gumbel-Morgenstern Distributions," *Communications in Statistics*, 415–427.

Johnson, N. L. and S. Kotz (1977). "On Some Generalized Farlie-Gumbel-Morgenstern Distribution–II Regression, Correlation and Further Generalizations," *Communications in Statistics*, **A6**, 485–496.

Johnson, N. L. and C. A. Rogers (1951). "The Moment Problem for Unimodal Distributions," *Annals of Mathematical Statistics*, **22**, 433–439.

Jones, M. C. (1985). "Generating Inverse Wishart Matrices," *Communications in Statistics—Simulation and Computation*, **14**, 511–514.

Kelker, D. (1970). "Distribution Theory of Spherical Distributions and a Location-Scale Parameter Generalization," *Sankyha*, 419–430.

Kennedy, W. J., Jr. and J. E. Gentle (1980). *Statistical Computing*. Marcel Dekker, New York.

Khintchine, A. Y. (1938). "On Unimodal Distributions," *Tomsk. Universitet. Nauchno-issledovatel'skii institut matematiki i mekhaniki IZVESITIIA*, **2**, 1–7.

Kimeldorf, G. and A. R. Sampson (1978). "Monotone Dependence," *Annals of Statistics*, **6**, 895–903.

Kinderman, A. J. and J. F. Monahan (1977). "Computer Generation of Random Variables Using the Ratio of Uniform Deviates," *ACM Transactions on Mathematical Software*, **3**, 257–260.

Kinderman, A. J. and J. G. Ramage (1976). "Computer Generation of Normal Random Variables," *Journal of the American Statistical Association*, **71**, 893–896.

Knuth, D. E. (1981). *The Art of Computer Programming, Seminumerical Algorithms*, 2nd ed. Addison-Wesley, Reading, Massachusetts.

Koehler, K. J. (1983). "A Simple Approximation for the Percentiles of the *t* Distribution," *Technometrics*, **25**, 103–105.

Koffler, S. L. and D. A. Penfield (1978). "Nonparametric Discrimination Procedures for Non-Normal Distributions," *Journal of Statistical Computation and Simulation*, **8**, 281–299.

Kotz, S. (1975). "Multivariate Distributions at a Cross Road," *Statistical Distributions in Scientific Work*, **1**, 247–270.

Kronmal, R. A. and A. V. Peterson, Jr., (1981). "A Variant of the Acceptance-Rejection Method for Computer Generation of Random Variables," *Journal of the American Statistical Association*, **76**, 446–451.

Krzanowski, W. J. (1977). "The Performance of Fisher's Linear Discriminant Function Under Non-Optimal Conditions," *Technometrics*, **19**, 191–200.

Lachenbruch, P. A., C. Sneeringer, and L. T. Revo (1973). "Robustness of the Linear and Quadratic Discriminant Function to Certain Types of Non-Normality," *Communications in Statistics*, **1**, 39–56.

Lai, C. D. (1978). "Morgenstern's Bivariate Distribution and its Application to Point Processes," *Journal of Mathematical Analysis and Applications*, **65**, 247–256.

Law, A. M. and W. D. Kelton (1982). *Simulation Modeling and Analysis*, McGraw-Hill, New York.

Lawrance, A. J. and P. A. W. Lewis (1983). "Simple Dependent Pairs of Exponential and Uniform Random Variables," *Operations Research*, 31, 1179–1197.

Lewis, T. G. and W. H. Payne (1973). "Generalized Feedback Shift Register Pseudo Random Number Algorithm," *Journal of the Association of Computing Machinery*, 20, 456–468.

Lord, R. D. (1954). "The Use of the Hankel Transform in Statistics, I. General Theory and Examples," *Biometrika*, 41, 44–55.

Mage, D. T. (1980). "An Explicit Solution for S_B Parameters Using Four Percentile Points," *Technometrics*, 22, 247–251.

Malkovich, J. F. and A. A. Afifi (1973). "On Tests for Multivariate Normality," *Journal of the American Statistical Association*, 68, 176–179.

Marasinghe, M. G. and W. J. Kennedy, Jr. (1982). "Direct Methods for Generating Extreme Characteristic Roots of Certain Random Matrices," *Communications in Statistics—Computation and Simulation*, 11, 527–542.

Mardia, K. V. (1967). "Some Contributions to Contingency-Type Distributions." *Biometrika*, 54, 235–249.

Mardia, K. V. (1969). "The Performance of Some Tests of Independence of Contingency-Type Bivariate Distributions," *Biometrika*, 56, 449–451.

Mardia, K. V. (1970). *Families of Bivariate Distributions*. Hafner, Darien, Connecticut.

Mardia, K. V. (1972). *Statistics of Directional Data*. Academic, New York.

Mardia, K. V., J. T. Kent, and J. M. Bibby (1979). *Multivariate Analysis*. Academic, New York.

Margolin, B. H. and L. W. Schruben (1978). "Pseudorandom Number Assignment in Statistically Designed Simulation and Distribution Sampling Experiments" (with discussion), *Journal of the American Statistical Association*, 73, 504–525.

Marsaglia, G. (1972). "Choosing a Point from the Surface of a Sphere," *The Annals of Mathematical Statistics*, 43, 645–646.

Marsaglia, G. and T. A. Bray (1964). "A Convenient Method for Generating Normal Variables," *SIAM Review*, 6, 260–264.

Marsalia, G., M. D. MacLaren, and T. A. Bray (1964). "A Fast Procedure for Generating Normal Random Variables," *Communications of the Association of Computing Machinery*, 7, 4–10.

Marsaglia, G. and W. W. Tsang (1984). "A Fast, Easily Implemented Method for Sampling from Decreasing or Symmetric Unimodal Density Functions," *SIAM Journal of Scientific and Statistical Computation*, 5, 349–359.

Martinez, J. and B. Iglewicz (1984). "Some Properties of the Tukey g and h Family of Distributions," *Communications in Statistics—Theory and Methods*, 13, 353–369.

McGraw, D. K. and J. F. Wagner (1968). "Elliptically Symmetric Distributions," *IEEE Transactions on Information Theory*, 14, 110–120.

Mihram, G. A. and R. A. Hultquist (1967). "A Bivariate Warning-Time/Failure-Time Distribution," *Journal of the American Statistical Association*, 62, 589–599.

Morgenstern, D. (1956). "Einfache Beispiele Zweidimensionaler Verteilungen," *Mitteilungsblatt für Mathematische Statistik*, 8, 234–235.

Mosteller, F. (1968). "Association and Estimation in Contingency Tables," *Journal of the American Statistical Association*, 63, 1–28.

Muirhead, R. J. (1982). *Aspects of Multivariate Statistical Theory*. John Wiley & Sons, New York.

Muller, M. E. (1959). "A Note on a Method for Generating Points Uniformly on N-Dimensional Spheres," *Communications of the ACM*, **2**, 19–20.

Nachtsheim, C. J. and M. E. Johnson (1986). "Some New Anisotropic Distributions Useful in Simulation," in preparation.

Nath, R. and B. S. Duran (1983). "A Robust Test in the Multivariate Two-Sample Location Problem," *American Journal of Mathematical and Management Sciences*, **3**, 225–249.

Odell, P. L. and A. H. Feiveson (1966). "A Numerical Procedure to Generate a Sample Covariance Matrix," *Journal of the American Statistical Association*, **61**, 199–203.

Olshen, R. and L. J. Savage (1970). "A Generalized Unimodality," *Journal of Applied Probability*, **7**, 21–34.

Patil, G. P., S. Kotz, and J. K. Ord, Eds. (1975). *A Modern Course on Statistical Distributions in Scientific Work*, Vol. 1. *Models and Structures*; Vol. 2. *Model Building and Model Selection*; Vol. 3. *Characterizations and Applications*. D. Reidel, Dordrecht, Holland/Boston, USA; The University of Calgary, Calgary, Alberta, Canada, July 29–August 10, 1974.

Patterson, D. A., F. L. Pirkle, M. E. Johnson, T. R. Bement, N. K. Stablein, and C. K. Jackson (1981). "Discriminant Analysis Applied to Aerial Radiometric Data and Its Application to Uranium Favorability in South Texas," *Journal of the International Association for Mathematical Geology*, **13**, 535–568.

Pearson, E. S., R. B. D'Agostino, and K. O. Bowman (1977). "Tests for Departures for Normality: Comparison of Powers," *Biometrika*, **64**, 321–346.

Plackett, R. L. (1965). "A Class of Bivariate Distributions," *Journal of the American Statistical Association*, **60**, 516–522.

Raftery, A. E. (1982). "Generalized Non-Normal Time Series Models," *Time Series Analysis: Theory and Practice, I* (O. D. Anderson, Ed.). North-Holland, Amsterdam, pp. 621–640.

Raftery, A. E. (1984). "A Continuous Multivariate Exponential Distribution," *Communications in Statistics—Theory and Methods*, **13**, 947–965.

Raftery, A. E. (1985). "Some Properties of a New Continuous Bivariate Exponential Distribution," *Statistics and Decisions*, **2**, 53–58.

Ramberg, J. S., P. R. Tadikamalla, E. J. Dudewicz, and E. F. Mykytka (1979). "A Probability Distribution and its Uses in Fitting Data," *Technometrics*, **21**, 201–214.

Ramshaw, L. A. and P. D. Amer (1982). "Random Number Generation Under Time and Space Constraints," Department of Computer and Information Sciences Technical Report, University of Delaware.

Rosenblatt, M. (1952). "Remarks on a Multivariate Transformation," *Annals of Mathematical Statistics*, **23**, 470–472.

Rubinstein, R. Y. (1981). *Simulation and the Monte Carlo Method*. John Wiley & Sons, New York.

Satterthwaite, S. P. and T. P. Hutchinson (1978). "A Generalisation of Gumbel's Bivariate Logistic Distribution," *Metrika*, **25**, 163–170.

Schmeiser, B. W. and R. Lal (1982). "Bivariate Gamma Random Vectors," *Operations Research*, **30**, 355–374.

Schmeiser, B. W. and M. A. Shalaby (1980). "Acceptance/Rejection Methods for Beta Variate Generation," *Journal of the American Statistical Association*, **75**, 673–678.

Schucany, W. R., W. C. Parr, and J. E. Boyer (1978). "Correlation Structure in Farlie-Gumbel-Morgenstern Distributions," *Biometrika*, **65**, 650–653.

Sibuya, M. (1962). "A Method for Generating Uniformly Distributed Points on n-Dimensional Spheres," *Annals of the Institute of Statistical Mathematics*, **14**, 81–85.

Silva, de B. M. (1978). "A Class of Multivariate Symmetric Stable Distributions," *Journal of Multivariate Analysis*, **8**, 335–345.

Slifker, J. F. and S. S. Shapiro (1980). "The Johnson System: Selection and Parameter Estimation," *Technometrics*, **22**, 239–246.

Smith, W. B. and R. R. Hocking (1972). "Wishart Variate Generation," *Applied Statistics*, **21**, 341–345.

Strassen, V. (1965). "The Existence of Probability Measures with Given Marginals," *Annals of Mathematical Statistics*, **36**, 423–439.

Subbotin, M. T. (1923). "On the Law of Frequency of Errors," *Mathematicheskii Sbornik*, **31**, 296–301.

Tadikamalla, P. R. R. (1980). "Random Sampling From the Exponential Power Distribution," *Journal of the American Statistical Association*, **75**, 683–686.

Tadikamalla, P. R. R. and M. E. Johnson (1981). "A Complete Guide to Gamma Variate Generation," *American Journal of Mathematical and Management Sciences*, **1**, 213–236.

Tadikamalla, P. R. R. and N. L. Johnson (1982a). "Systems of Frequency Curves Generated by Transformation of Logistic Variables," *Biometrika*, **69**, 461–465.

Tadikamalla, P. R. R. and N. L. Johnson (1982b). "Tables to Facilitate Fitting L_U Distributions," *Communications in Statistics—Computation and Simulation*, **11**, 249–271.

Taillie, C., G. P. Patil, and B. A. Baldessari, Eds. (1981). *Statistical Distributions in Scientific Work*, Vol. 4. *Models, Structures, and Characterizations*; Vol. 5. *Inferential Problems and Properties*; Vol. 6. *Applications in Physical, Social, and Life Sciences*. D. Reidel, Dordrecht, Holland/Boston, USA.

Takahasi, K. (1965). "Note on the Multivariate Burr's Distribution," *Annals of the Institute of Statistical Mathematics*, **17**, 257–260.

Tashiro, Y. (1977). "On Methods for Generating Uniform Points on the Surface of a Sphere," *Annals of the Institute of Statistical Mathematics*, **29**, 295–300.

Watson, G. S. (1983). *Statistics on Spheres*. John Wiley & Sons, New York.

Whitt, W. (1976). "Bivariate Distributions with Given Marginals," *Annals of Statistics*, **4**, 1280–1289.

Wilson, J. R. (1983). "Antithetic Sampling with Multivariate Inputs," *American Journal of Mathematical and Management Sciences*, **3**, 121–144.

Wilson, J. R. (1984). "Variance Reduction Techniques for Digital Simulation," *American Journal of Mathematical and Management Sciences*, **4**, 277–312.

Wolfe, S. J. (1975). "On the Unimodality of Spherically Symmetric Stable Distribution Functions," *Journal of Multivariate Analysis*, **5**, 236–242.

Supplementary References

Abrahams, J. and J. B. Thomas (1984). "A Note on the Characterization of Bivariate Densities by Conditional Densities," *Communications in Statistics—Theory and Methods*, **13**, 395–400.

Anderson, D. A. (1977). "Statistical Inference for a Certain Class of Bivariate Distributions," *Journal of Multivariate Analysis*, **7**, 560–571.

Andrews, D. F. and C. L. Mallows (1974). "Scale Mixtures of Normal Distributions," *Journal of the Royal Statistical Society*, Series B, **36**, 99–102.

Athreya, K. B. (1986). "Another Conjugate Family for the Normal Distribution," *Statistics and Probability Letters*, **4**, 61–64.

Barndorff-Nielsen, O. and P. Blaesild (1981). "Hyperbolic Distributions and Ramifications: Contributions to Theory and Application," *Statistical Distributions in Scientific Work*, **4**, 19–44.

Barnett, V. (1980). "Some Bivariate Uniform Distributions," *Communications in Statistics—Theory and Methods*, **9**, 339–353.

Basu, A. P. and H. W. Block (1975). "On Characterizing Univariate and Multivariate Exponential Distributions with Applications," *Statistical Distributions in Scientific Work*, **3**, 399–421.

Beran, R. (1979). "Testing for Ellipsoidal Symmetry of a Multivariate Density," *Annals of Statistics*, **7**, 150–162.

Bertin, E. M. J. and R. Theodorescu (1980). "Some Characterizations of Discrete Unimodality," *Statistics and Probability Letters*, **2**, 23–30.

Blaesild, P. and J. L. Jensen (1981). "Multivariate Distributions of Hyperbolic Type," *Statistical Distributions in Scientific Work*, **4**, 45–66.

Blake, I. F. and J. B. Thomas (1968). "On a Class of Processes Arising in Linear Estimation Theory," *IEEE Transactions on Information Theory*, **14**, 12–16.

Brandwein, A. C. (1979). "Minimax Estimation of the Mean of Spherically Symmetric Distributions under General Quadratic Loss," *Journal of Multivariate Analysis*, **9**, 579–588.

Cambanis, S. (1983). "Complex Symmetric Stable Variables and Processes," *Contributions to Statistics: Essays in Honour of Norman L. Johnson* (P. K. Sen, Ed.). North-Holland, Amsterdam.

Chambers, J. M. (1970). "Computers in Statistical Research: Simulation and Computer-Aided Mathematics," *Technometrics*, **12**, 1–15.

Cheng, R. C. H. (1985). "Generation of Multivariate Normal Samples with Given Sample Mean and Covariance Matrix," *Journal of Statistical Computation and Simulation*, **21**, 39–49.

Cheng, R. C. H. and G. M. Feast (1980). "Gamma Variate Generators with Increased Shape Parameter Range," *Communications of the ACM*, **23**, 389–394.

Chmielewski, M. A. (1980). "Invariant Scale Matrix Hypothesis Tests Under Elliptical Symmetry," *Journal of Multivariate Analysis*, **10**, 343–350.

Chmielewski, M. A. (1981). "Invariant Tests for the Equality of *k* Scale Parameters Under Spherical Symmetry," *Journal of Statistical Planning and Inference*, **5**, 341–346.

Conway, D. (1981). "Bivariate Distribution Contours," *Proceedings of the Business and Economic Statistics Section*, 475–480.

Cuadras, C. M. and J. Auge (1981). "A Continuous General Multivariate Distribution and its Properties," *Communications in Statistics—Theory and Methods*, Series A, **10**, 339–353.

Devlin, S. J., R. Gnanadesikan, and J. R. Kettenring (1981). "Robust Estimation of Dispersion Matrices and Principal Components," *Journal of the American Statistical Association*, **76**, 354–362.

Devroye, L. (1984). "Methods for Generating Random Variates with Polya Characteristic Functions," *Statistics and Probability Letters*, **2**, 257–261.

Devroye, L. (1984). "A Simple Algorithm for Generating Random Variates with a Log-Concave Density, *Computing*, **33**, 247–257.

Devroye, L. (1986). *Non-Uniform Random Variate Generation*. Springer-Verlag, New York.

Dietz, E. J. (1982). "Bivariate Nonparametric Tests for the One-Sample Location Problem," *Journal of the American Statistical Association*, **77**, 163–169.

Diggle, P. J., N. I. Fisher, and A. J. Lee (1985). "A Comparison of Tests of Uniformity for Spherical Data," *Australian Journal of Statistics*, **27**, 53–59.

Dudewicz, E. J. (1975). "Random Numbers: The Need, The History, The Generators," *Statistical Distributions in Scientific Work*, **2**, 25–36.

Dussauchoy, A. and R. Berland (1975). "A Multivariate Gamma Type Distribution Whose Marginal Laws Are Gamma, and which Has a Property Similar to a Characteristic Property of the Normal Case," *Statistical Distributions in Scientific Work*, **1**, 319–328.

Ebrahimi, N. and M. Ghosh (1981). "Multivariate Negative Dependence," *Communications in Statistics—Theory and Methods*, **10**, 307–337.

Efron, B. (1982). "Transformation Theory: How Normal Is a Family of Distributions?," *The Annals of Statistics*, **10**, 323–339.

Franklin, J. N. (1965). "Numerical Simulation of Stationary and Non-Stationary Gaussian Random Processes," *SIAM Review*, **7**, 68–80.

Friday, D. S. (1981). "Dependence Concepts for Stochastic Processes," *Statistical Distributions in Scientific Work*, **5**, 349–361.

Gaver, D. P. (1970). "Multivariate Gamma Distributions Generated by Mixture," *Sankhya*, Series A, **32**, 123–126.

Gaver, D. P. and M. Acar (1979). "Analytical Hazard Representations for Use in Reliability, Mortality, and Simulation Studies," *Communications in Statistics—Simulation and Computation*, Series A, **8**, 91–111.

Gaver, D. P. and P. A. W. Lewis (1980). "First-Order Autoregressive Gamma Sequences and Point Processes," *Advances in Applied Probability*, **12**, 727–745.

Genest, C. and R. J. MacKay (1985). "Archimedean Copulas and Families of Bivariate Distributions," unpublished manuscript.

Ghosh, M. and E. Pollak (1975). "Some Properties of Multivariate Distributions with PDF's Constant on Ellipsoids," *Communications in Statistics*, **4**, 1157–1160.

Gine, E. and M. G. Hahn (1983). "On Stability of Probability Laws with Univariate Stable Marginals," *Zeitschrift für Wahrscheinlichkeitstheorie*, **64**, 157–165.

Goodman, I. R. and S. Kotz (1981). "Hazard Rates Bases on Isoprobability Contours," *Statistical Distributions in Scientific Work*, **5**, 289–308.

Gumbel, E. J. (1961). "Multivariate Exponential Distributions," *Bulletin of the International Statistical Institute*, **39**, 469–475.

Gupta, A. K. and C. F. Wong (1985). "On Three and Five Parameter Bivariate Beta Distributions," *Metrika*, **32**, 85–91.

Gupta, R. P. (1975). "Multivariate Beta Distribution," *Statistical Distributions in Scientific Work*, **1**, 337–344.

Haralick, R. M. (1977). "Pattern Discrimination Using Ellipsoidally Symmetric Multivariate Density Functions," *Pattern Recognition*, **9**, 89–94.

Huster, W. J. and R. Brookmeyer (1986). "The Diabetic Retinopathy Study Revisited," unpublished manuscript.

Hutchinson, T. P. (1981). "Compound Gamma Bivariate Distributions," *Metrika*, **28**, 263–271.

Javier, W. R. and A. K. Gupta (1985). "On Matric Variate-*t* Distribution," *Communications in Statistics—Theory and Methods*, **14**, 1413–1425.

Jensen, D. R. (1984). "The Structure of Ellipsoidal Distributions I. Canonical Analysis," *Biometrical Journal*, **26**, 779–787.

Jensen, D. R. and I. J. Good (1983). "A Representation for Ellipsoidal Distributions," *SIAM Journal of Applied Mathematics*, **43**, 1194–1200.

Jogdeo, K. (1975). "Dependence Concepts and Probability Inequalities," *Statistical Distributions in Scientific Work*, **1**, 271–279.

Johnson, N. L. (1980). "Extreme Sample Censoring Problems with Multivariate Data: Indirect Censoring and the Farlie-Gumbel-Morgenstern Distribution," *Journal of Multivariate Analysis*, **10**, 351–362.

Johnson, N. L. and S. Kotz (1975). "A Vector Multivariate Hazard Rate," *Journal of Multivariate Analysis*, **5**, 53–66.

Jones, R. M. and K. S. Miller (1966). "On the Multivariate Lognormal Distribution," *Journal of the Industrial Mathematics Society*, **16**, 63–76.

Kafaei, M.-A. and P. Schmidt (1985). "On the Adequacy of the 'Sargan Distribution' as an Approximation to the Normal," *Communications in Statistics—Theory and Methods*, **14**, 509–526.

Kimeldorf, G. and A. Sampson (1975). "Uniform Representations of Bivariate Distributions," *Communications in Statistics*, **4**, 617–627.

Kimeldorf, G. and A. Sampson (1975). "One-Parameter Families of Bivariate Distributions with Fixed Marginals," *Communications in Statistics*, **4**, 293–301.

Kinderman, A. J., J. F. Monahan, and J. G. Ramage (1977). "Computer Methods for Sampling from Student's *t* Distributions," *Mathematics of Computation*, **31**, 1009–1018.

Kingman, J. F. C. (1972). "On Random Sequences with Spherical Symmetry," *Biometrika*, **59**, 492–494.

Kowalski, C. J. (1973). "Non-Normal Bivariate Distributions with Normal Marginals," *The American Statistician*, **27**, 103–106.

Kshirsagar, A. M. (1959). "Bartlett Decomposition and Wishart Distribution," *Annals of Mathematical Statistics*, **30**, 239–241.

Lawrance, A. J. and P. A. W. Lewis (1980). "The Exponential Autoregressive-Moving Average EARMA(p, q) Process," *Journal of the Royal Statistical Society*, Series B, **42**, 150–161.

Lawrance, A. J. and P. A. W. Lewis (1981). "A New Autoregressive Time Series Model in Exponential Variables (NEAR(1))," *Advances in Applied Probability*, **13**, 826–845.

Lawrance, A. J. and P. A. W. Lewis (1982). "Generation of Some First-Order Autoregressive Markovian Sequences of Positive Random Variables with Given Marginal Distributions," *Applied Probability-Computer Science: The Interface Volume 1* (R. L. Disney and T. J. Ott, Eds.). Birkhauser, Boston.

Lawrance, A. J. and P. A. W. Lewis (1985). "Modelling and Residual Analysis of Nonlinear Autoregressive Time Series in Exponential Variables," *Journal of the Royal Statistical Society*, Series B, **47**, 165–202.

Lee, L. (1979). "Multivariate Distributions Having Weibull Properties," *Journal of Multivariate Analysis*, **9**, 267–277.

Lin, G. D. (1986). "Relationships Between Two Extensions of Farlie-Gumbel-Morgenstern Distribution," unpublished manuscript.

Mallows, C. L. (1983). "A New System of Frequency Curves," *Contributions to Statistics: Essays in Honour of Norman L. Johnson*, (P. K. Sen, Ed.). North-Holland, Amsterdam.

Mardia, K. V. (1981). "Recent Directional Distributions with Applications," *Statistical Distributions in Scientific Work*, **6**, 1–19.

Marsaglia, G. (1964). "Generating a Variable from the Tail of the Normal Distribution," *Technometrics*, **6**, 101–102.

Marsaglia, G. (1965). "Ratios of Normal Variables and Ratios of Sums of Uniform Variables," *Journal of the American Statistical Association*, **60**, 193–204.

Marsaglia, G. (1977). "The Squeeze Method for Generating Gamma Variates," *Computers and Mathematics with Applications*, **3**, 321–325.

Marsaglia, G. and I. Olkin (1984). "Generating Correlation Matrices," *SIAM Journal of Scientific and Statistical Computing*, **5**, 470–475.

Marshall, A. W. and I. Olkin (1967). "A Multivariate Exponential Distribution," *Journal of the American Statistical Association*, **62**, 30–44.

Marshall, A. W. and I. Olkin (1985). "A Family of Bivariate Distributions Generated by the Bivariate Bernoulli Distribution," *Journal of the American Statistical Association*, **80**, 332–338.

Mathar, R. (1985). "Outlier-Prone and Outlier-Resistant Multidimensional Distributions," *Statistics*, **16**, 451–456.

Maxwell, W. L. (1972). "Generation of Multivariate and Serially Correlated Random Variables," Cornell Technical Report No. 149.

McKenzie, E. (1985). "An Autoregressive Process for Beta Random Variables," *Management Science*, **31**, 988–997.

McNolty, F., J. R. Hugnen, and E. Hansen (1975). "Certain Statistical Distributions Involving Special Functions and their Applications," *Statistical Distributions in Scientific Work*, **1**, 131–160.

Miller, K. S. (1983). "Some Skew Multivariate Distributions Which Follow a Law," *Journal of the Industrial Mathematics Society*, **33**, 127–138.

Moran, P. A. P. (1967). "Testing for Correlation Between Non-Negative Variates," *Biometrika*, **54**, 385–394.

Moran, P. A. P. (1967). "Testing for Serial Correlation with Exponentially Distributed Variates," *Biometrika*, **54**, 395–401.

Nataf, A. (1962). "Determination des Distributions de Probabilities dont des Marges sont Donnees," *Comptes Rendus*, **255**, 43–44.

Oliveira, J. Tiago de (1975). "Bivariate and Multivariate Extreme Distributions," *Statistical Distributions in Scientific Work*, **1**, 355–361.

Pillai, R. N. (1985). "Semi-α-Laplace Distributions," *Communications in Statistics—Theory and Methods*, **14**, 991–1000.

Prentice, R. L. (1974). "A Log Gamma Model and Its Maximum Likelihood Estimation," *Biometrika*, **61**, 539–544.

Randles, R. H., J. D. Broffitt, J. S. Ramberg, and R. V. Hogg (1978). "Discriminant Analysis Based on Ranks," *Journal of the American Statistical Association*, **73**, 379–384.

Ratnaparkhi, M. V. (1981). "Some Bivariate Distributions of (X, Y) Where the Conditional Distribution of Y, Given X, Is Either Beta or Unit-Gamma," *Statistical Distributions in Scientific Work*, **4**, 389–400.

Richards, D. (1984). "Hyperspherical Models, Fractional Derivatives and Exponential Distributions on Matrix Spaces," *Sankhya*, Series A, **46**, 155–165.

Richards, D. St. P. (1986). "Positive Definite Symmetric Functions on Finite Dimensional Spaces, I: Applications of the Radon Transform," submitted.

Rodriguez, R. N. (1977). "A Guide to the Burr Type XII Distributions," *Biometrika*, **64**, 129–134.

Rogers, W. H. and J. W. Tukey (1972). "Understanding Some Long-Tailed Symmetrical Distributions," *Statistica Neerlandica*, **26**, 211–226.

Roux, J. J. J. (1981). "New Families of Multivariate Distributions," *Statistical Distributions in Scientific Work*, **1**, 281–297.

Roux, J. J. J. and P. J. Becker (1981). "Compound Distributions Relevant to Life Testing," *Statistical Distributions in Scientific Work*, **4**, 111–124.

Rubin, P. A. (1984). "Generating Random Points in a Polytope," *Communications in Statistics—Simulation and Computation*, **13**, 375–396.

Scheuer, E. M. and D. S. Stoller (1962). "On the Generation of Normal Random Vectors," *Technometrics*, **4**, 278–281.

Schmeiser, B. and R. Lal (1980). "Multivariate Modeling in Simulation: A Survey," *ASQC Technical Conference Transactions*, *Atlanta*, American Society for Quality Control, Milwaukee, pp. 252–261.

Schmeiser, B. and R. Lal (1980). "Squeeze Methods for Generating Gamma Variates," *Journal of the American Statistical Association*, **75**, 679–682.

Schoenberg, I. J. (1938). "Metric Spaces and Completely Monotone Functions," *Annals of Mathematics*, **39**, 811–841.

Shaked, M. (1975). "A Note on the Exchangeable Generalized Farlie-Gumbel-Morgenstern Distributions," *Communications in Statistics*, **4**, 711–721.

Shanthikumar, J. G. (1985). "Discrete Random Variate Generation Using Uniformization," *European Journal of Operational Research*, **21**, 387–398.

Simkowitz, M. A. and W. L. Beedles (1980). "Asymmetric Stable Distributed Security Returns," *Journal of the American Statistical Association*, **75**, 306–312.

Small, N. J. H. (1980). "Marginal Skewness and Kurtosis in Testing Multivariate Normality," *Applied Statistics*, **29**, 85–87.

Stadje, W. (1984). "On the Theory of Rotationally Symmetric Random Vectors," *Statistics and Decisions*, **2**, 175–193.

Taillie, C. (1981). "Lorenz Ordering Within the Generalized Gamma Family of Income Distributions," *Statistical Distributions in Scientific Work*, **6**, 181–192.

Tamura, R. N. and D. D. Boos (1986). "Minimum Hellinger Distance Estimation for Multivariate Location and Covariance," *Journal of the American Statistical Association*, **81**, 223–229.

Tubbs, J. D. and O. E. Smith (1985). "A Note on the Ratio of Positively Correlated Gamma Variates," *Communications in Statistics—Theory and Methods*, **14**, 13–23.

Wahrendorf, J. (1980). "Inference in Contingency Tables with Ordered Categories Using Plackett's Coefficient of Association for Bivariate Distributions," *Biometrika*, **67**, 15–21.

Author Index

Subject Index

(continued from front)